RESTRICTED
FILE: B

TECHNICAL ORDER NO. 01-20EF-1

B-17F
AIRPLANE

PILOT'S FLIGHT OPERATING INSTRUCTIONS

NOTICE: This document contains information affecting the National Defense of the United States within the meaning of the Espionage Act, 50 U. S. C., 31 and 32, as amended. Its transmission or the revelation of its contents in any manner to an unauthorized person is prohibited by law.

©2006-2009 PERISCOPE FILM LLC
ALL RIGHTS RESERVED
ISBN #978-1-4116-8725-7

This manual is sold for historic research purposes only, as an entertainment. It is not intended to be used as part of an actual flight training program. No book can substitute for flight training by an authorized instructor. The licensing of pilots is overseen by organizations and authorities such as the FAA and CAA. Operating an aircraft without the proper license is a federal crime.

PUBLISHED BY AUTHORITY OF THE COMMANDING GENERAL,
ARMY AIR FORCES, BY THE HEADQUARTERS,
AIR SERVICE COMMAND, PATTERSON FIELD, FAIRFIELD, OHIO

The A. L. Garber Co., Ashland, O.—4-43, 22,000

DECEMBER 25, 1942
Revised 4-1-43

LIST OF REVISED PAGES ISSUED

Page	Latest Revised Date
1	4-1-43
8	4-1-43
13	4-1-43
25	4-1-43
26	4-1-43
50	4-1-43
54	4-1-43
58	4-1-43
60	4-1-43
62	4-1-43
64	4-1-43
68	4-1-43
69	4-1-43
108A	4-1-43
108B	4-1-43
145	4-1-43
148	4-1-43

NOTE: A heavy black vertical line, to the left of the text on revised pages, indicates the extent of the revision. This is omitted where more than 50 percent of the page is revised.

THIS PUBLICATION MAY BE USED BY PERSONNEL RENDERING SERVICE TO THE UNITED STATES OR ITS ALLIES

Paragraph 5.*d.* of Army Regulation 380-5 relative to the handling of "restricted" printed matter is quoted below.

"*d. Dissemination of restricted matter.*—The information contained in restricted documents and the essential characteristics of restricted material may be given to *any person known to be in the service of the United States and to persons of undoubted loyalty and discretion who are cooperating in Government work*, but will not be communicated to the public or to the press except by authorized military public relations agencies."

This permits the issue of "restricted" publications to civilian contract and other accredited schools engaged in training personnel for Government work, to civilian concerns contracting for overhaul and repair of aircraft or aircraft accessories, and to similar commercial organizations.

TABLE OF CONTENTS

Section		Page
I	Description	1- 47
	1. Airplane	1
	2. Power Plant	1
	3. Propeller	1
	4. Controls and Operational Equipment	1
II	Pilot's Compartment Instructions	48- 69
	1. Before Entering the Pilot's Compartment	48
	2. On Entering the Pilot's Compartment	48
	3. Starting the Engines	50
	4. Engine Warm-up	54
	5. Emergency Take-off	54
	6. Engine and Accessories Ground Test	54
	7. Taxiing	56
	8. Take-off	58
	9. Engine Failure During Take-off	58
	10. Climb	60
	11. Flight Operations	60
	12. Fire in Flight	62
	13. Propeller Feathering	62
	14. Propeller Unfeathering	64
	15. General Flying Characteristics	65
	16. Stalls	66
	17. Spins	67
	18. Dives	67
	19. Approach in Landing	67
	20. Emergency Take-off if Landing is Not Completed	68
	21. After Landing	68
	22. Stopping of Engines	69
	23. Before Leaving the Pilot's Compartment	69
	24. Maneuvers Prohibited	69
III	Flight Operation Data	70-89
	1. Determining Gross Weight	70
	2. Flight Planning	70
IV	Operating Instructions - Navigator's Compartment	93
	1. General Description	93
	2. Operational Equipment	93
V	Operating Instructions - Radio Compartment	95-102
	1. General Description	95
	2. Operational Equipment	95
VI	Operating Instructions - Bombardier's Compartment	103-108
	1. General Description	103
	2. Operational Equipment	103
VII	Operating Instructions - Upper Turret Compartment	112-118
	1. General Description	112
	2. Operational Equipment	112

RESTRICTED

TABLE OF CONTENTS

Section		Page
VIII	Operating Instructions - Ball Turret Compartment	119-125
	1. General Description	119
	2. Operational Equipment	119
IX	Operating Instructions - Side Gunner's Compartment	126-130
	1. General Description	126
	2. Operational Equipment	126
X	Operating Instructions - Bomb Bay Compartment	131-134
	1. General Description	131
	2. Operational Equipment	131
XI	Operating Instructions - Tail Gunner's Compartment	135-137
	1. General Description	135
	2. Operational Equipment	135
XII	Operating Instructions - Tail Wheel Compartment	138
	1. General Description	138
	2. Operational Equipment	138
XIII	Operating Instructions - Camera Pit	140
	1. General Description	140
	2. Operational Equipment	140
XIV	Cold Weather Operation	140-145
	1. Winterization	140
	2. Engine Oil Dilution System	141
	3. Propeller Oil Dilution	141
	4. Portable Ground Heaters	141
	5. Cold Weather Starting of Engines	143
	6. Batteries	143
	7. Protective Covers	145
	8. Frost or Ice Removal	145
	9. Mooring	145
	10. Communications Equipment	145
	11. Carburetor De-icing	145

Appendix

I	U.S.A. - British Glossary of Nomenclature	147
II	Emergency Operating Instructions	148-157
	1. Emergency Operation of Landing Gear	148
	2. Emergency Operation of the Tail Wheel	148
	3. Emergency Operation of Bomb Bay Doors	148
	4. Emergency Operation of Wing Flaps	148
	5. Emergency Bomb Release	148
	6. Fire Extinguishing Equipment	148
	7. Life Raft and First Aid	148
	8. Emergency Exits	151
	9. Alarm Bells	151
	10. Emergency Operation of Radio Equipment	151
	11. Suggested Methods of Abandoning Airplane	153
	12. Emergency Brake Operation	156

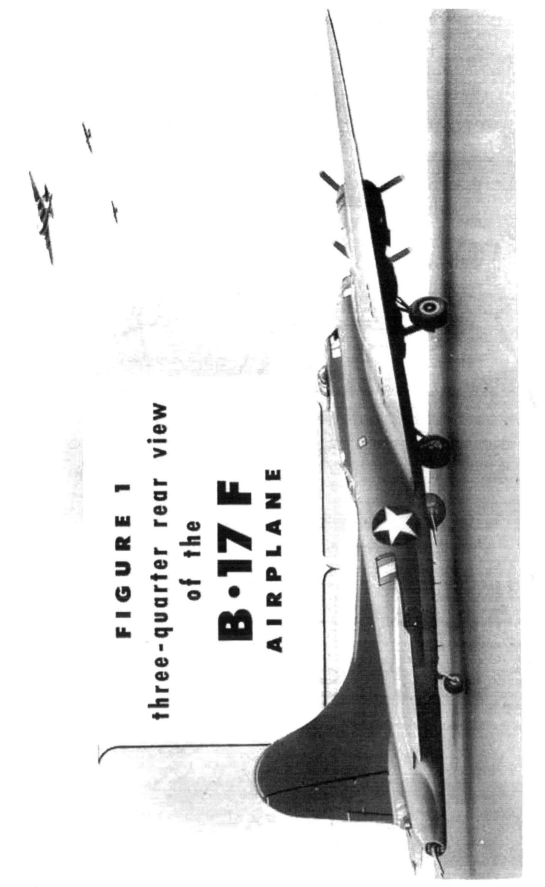

FIGURE 1 three-quarter rear view of the B-17F AIRPLANE

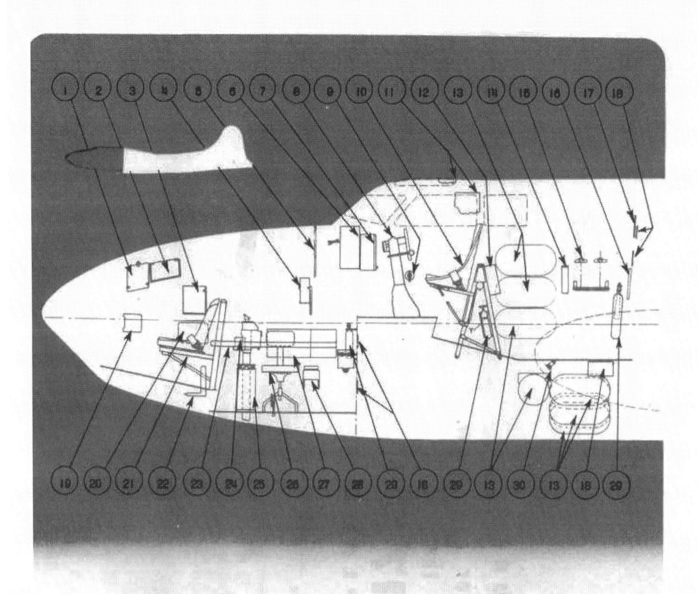

1	Clip Board	11	Curtains	21	Bombardier's Seat
2	Bomb Data Case	12	Wiring Diagram Box	22	Pilot Static Tube
3	Map Case	13	Oxygen Bottles	23	Navigator's Table
4	Navigator's Map Case	14	Cup Dispenser	24	Ash Receiver
5	Navigation Case	15	Thermos Bracket	25	Drift Meter
6	Instrument Panel	16	Flight Report Holder	26	Navigator's Seat
7	Copilot's Auxiliary Panel	17	Diagram Frame	27	Bomb Sight Storage Box
8	Control Column	18	Data Card Frame	28	Aperiodic Compass
9	Vacuum Control	19	Bomb Chart Plate	29	Fire Extinguisher
10	Pilot's Seat	20	Bombardier's Instrument Panel	30	Thermometer
		21	Bombardier's Seat		
		22	Pilot Static Tube		

Figure 2 – Fuselage Contents Arrangement Diagram (Sheet 1)

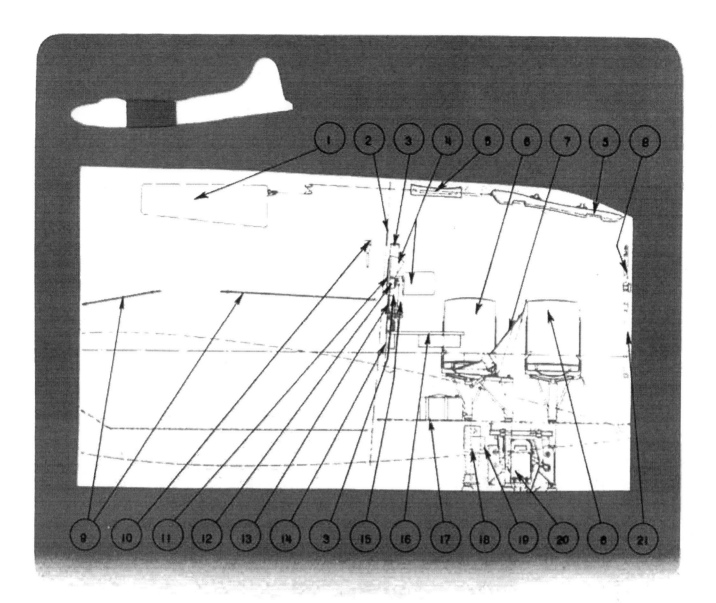

1 Life Raft	7 Radio Operator's Seat	16 Radio Operator's Table
2 Data Card	8 Handcrank And Extension	17 Shutter Induction Coil
3 Cup Dispenser	9 Rope Guard Rail	18 View Finder
4 Radio Table Rack	10 Relief Tube	19 Intervalometer
5 Black Out Curtain	11 Fuse Location Chart	20 Camera
6 Seats For Auxilliary Crew	12 Ash Receiver	21 Starter Crank Extension Support
	13 Thermos Bracket	
	14 Fire Extinguisher	
	15 Thermos Bottle	

Figure 2 - Fuselage Contents Arrangement Diagram (Sheet 2)

RESTRICTED

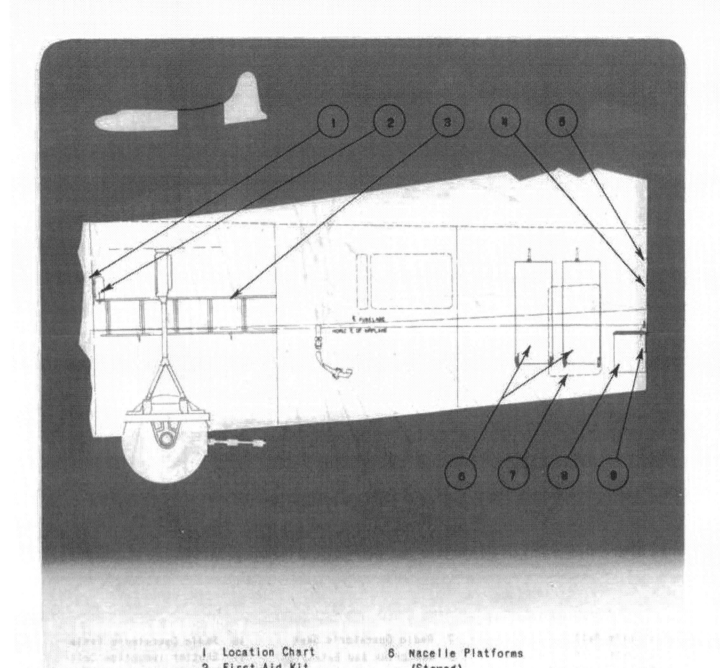

1 Location Chart
2 First Aid Kit
3 Ladder
4 Name Plate
5 Coat Hook
6 Nacelle Platforms (Stowed)
7 Main Entrance Door
8 Chemical Toilet
9 Fire Extinguisher

Figure 2 — Fuselage Contents Arrangement Diagram (Sheet 3)

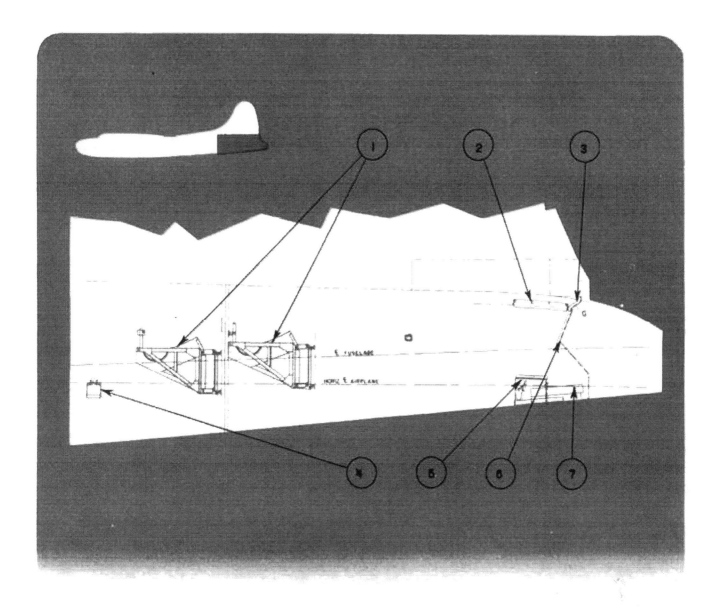

1. Nacelle Platform Hanger & Brace (Stowed)
2. Shoulder Pad
3. Chin Pad
4. Paper Holder
5. Tail Gunner's Seat
6. Chest Pad
7. Knee Pad

Figure 2 - Fuselage Contents Arrangement Diagram (Sheet 4)

THIS PAGE INTENTIONALLY LEFT BLANK

SECTION I

DESCRIPTION

1. **Airplane.**

    a. **General.** - The model B-17F bombardment airplane is a midwing monoplane, powered by four model R-1820-97 engines. Electrically-operated landing gear, tail gear, wing flaps and bomb bay doors, and hydraulically-operated brakes and cowl flaps are provided. This model airplane is equipped with automatic flight control equipment. The crew consists of pilot, copilot, navigator, bombardier, upper turret gunner, lower turret gunner, radio operator, side gunner, and tail gunner. The approximate overall dimensions are as follows:

Length	74 ft 8.90 in.
Height, taxiing position	19 ft 1.00 in.
Span	103 ft 9.38 in.

    b. **Access to the Airplane.** - The airplane can be entered through either the main entrance door located on the right side of the airplane just forward of the horizontal stabilizer or through the front entrance door located in the bottom of the fuselage below the pilot's compartment.

    c. **Mooring Instructions.** - Lines may be attached at towing points and also at the rope well located in the bottom surface of each wing. The rope wells are designed for a 10,000-pound load acting downward. Moor the airplane with the nose into the wind, set the parking brakes and lock the rudder and elevators. When attaching the mooring lines at the rope wells in the wings, allow approximately 16 inches slack in the line. This will prevent damage to the structure or loss of mooring control in case a tire goes flat with resultant elevation of the opposite wing. Rudder and elevator locks will withstand gust loads from any direction up to 60 mph wind velocity. (See figure 3.)

Figure 3 - Mooring Provisions Diagram

    d. **Hydraulic System.** (See figure 7.)

    e. **Oil System.** (See figure 8.)

    f. **Fuel System.** (See figure 9.)

    g. **Heating and Ventilating System.** (See figures 10, 11, and 12.)

    h. **Oxygen System.** (See figure 13.)

    i. **Armament.** - The pilot and copilot are protected by armor plate behind their seats. (See figure 14.) Refer to the section for each compartment for further detailed information, including protective armor plating and angles of armor protection.

2. **Power Plant.**

    a. The Wright Model R-1820-97 engine is an air-cooled nine cylinder radial aircraft engine, equipped with integral reduction gears through which the propeller is driven. The airplane installation includes a turbo-supercharger. The engine is equipped with a Bendix-Stromberg injection type PD12H3 carburetor.

    b. **Fuel and Oil.**

    (1) Fuel: 100 Octane;
        Specification: AN-VV-F-781

    (2) Oil: Specification: AN-VV-O-446
        Viscosity: 1120

    c. **Automatic Engine Control.** - Should engine control cables be shot away, four of the controls will automatically assume a predetermined position as follows:

Throttles	Wide Open
Supercharger Regulator	65% Power
Carburetor Air Temp. (intercooler)	Cold
Propeller Pitch	1850 rpm

    These settings are accomplished by spring action at each unit and functioning of the automatic control at one unit will not affect placement of controls at other units or of similar controls on other engines.

3. **Propeller.**

    The Hamilton Standard propeller, is equipped with three blades and is hydromatically controlled, with constant speed and full feathering provisions.

4. **Controls and Operational Equipment.**

    a. **Trim Tabs.** (See figure 21.)

    (1) The aileron trim tab knob controls the tab by cable operation of the actuating screw. Complete tab travel requires approximately 3-3/4 turns of the knob.

(2) The rudder trim tab wheel controls the tab by cable operation of the actuating screw. Complete tab travel requires approximately seven turns of the wheel.

(3) The two elevator trim tab wheels are on opposite sides of the control stand, and operate on the same shaft to control the tabs by cable operation of actuating screws. Complete tab travel requires approximately six turns of the wheel. A knurled thumb nut on the left side of the control stand provides a friction brake to prevent creeping of the tab control.

b. Locks. (See figure 22.)

(1) Rudder and Elevator. - The rudder and elevator locking lever (see inset) operates by cable control to place a pin in a socket on a segment at each of the control quadrants. The locking lever, which is recessed into the floor aft of the engine control stand, is locked in either the up or down position. The lever may be moved to the up, or locked position, regardless of the attitude of the control surfaces. Under this condition, the control surfaces will automatically lock when the rudder is in the neutral position and the elevator is in the down position.

(2) Aileron. - The aileron is locked by a pin which is manually inserted in a hole in the left control column and holds the center spoke of that wheel in a padded slot. The pin is clipped to the pilot's control column when not in use. The aileron locks in the neutral position.

(3) Tail Wheel. - The tail wheel locking lever (see inset) operates a single cable to retract a spring-loaded locking pin from a socket in the treadle. The locking lever which is recessed into the floor aft of the control stand, latches in the UP position only and may be moved into the down position regardless of the attitude of the tail wheel, which will lock when centered. To release the locking handle, press the knob on the end of it. A red signal light on the pilot's instrument panel is off when the tail wheel is locked and is controlled through a switch which is operated by motion of the locking pin.

c. Carburetor Temperature Controls. - The shutters on the intercooler are cable-controlled from a stand at the right side wall in front of the copilot. Each of the four cables is operated by a slide which latches in any desired position throughout the range from full open to full closed. To release the latch, pull out on the handle.

d. Propeller Feathering and Unfeathering.

(1) Each propeller is controlled individually by one of the four magnetic push button switches (see inset) located at the bottom of the instrument panel. Each switch controls a solenoid switch in the corresponding nacelle which operates an electric motor-driven hydraulic pump. Pressure in the system increases rapidly and at 400 pounds per square inch, a pressure cut-out switch on the propeller control head opens the push button holding coil circuit, which releases the motor control and the pump stops. If it is necessary to stop the feathering operating before it is completed, the switch may be pulled open by hand.

(2) In order to unfeather the propeller, the push button switch must be manually held in the closed position until unfeathering has been accomplished.

NOTE: When unfeathering a propeller on a cold engine, do not allow the engine speed to exceed minimum governing speed until oil pressure and oil temperature appear satisfactory. Turn off the ignition after feathering any propeller if the engine is to remain inoperative for any appreciable length of time. Do not operate more than one propeller feathering switch at a time except in emergencies.

e. __Primer.__ - The cylinder head primer (see inset) has four positions corresponding to each of the four engines and an "OFF" position. The primer handle is locked only in the "OFF" position. To operate, push the handle down, turn the valve to the engine position required and then withdraw the handle and pump the charge to the cylinder head in the conventional manner.

IMPORTANT: It should be noted, however, that pressure from No. 3 fuel booster pump is on the suction side of the primer and over priming will result if the handle is left in the withdrawn position. Therefore, each priming operation must terminate with the handle returned to the locked position.

f. __Parking Brake.__ - The pull handle at the bottom of the instrument panel will set the copilot's brake metering valves when the foot pedals are depressed. This utilizes the regular braking system, and therefore, hydraulic system pressure must be available when the parking brake is required for any length of time. When necessary, set the parking brake handle and pump the system pressure to at least 400 pounds per square inch. This is the minimum pressure at which full braking control is available.

WARNING: Do not set parking brake for a long period of time if brake drums are hot.

g. __Engine Fire Extinguisher.__

(1) The selector valve on the auxiliary panel in front of the copilot may be turned to any one of the four engine positions, in order to direct the discharge from the CO_2 cylinder to the desired engine extinguisher ring.

(2) Two pull handles one on either side of the selector valve, are provided to control the discharge of the two CO_2 cylinders located in the right wing gap forward of the rear spar. Do not attempt to distribute the discharge from one bottle to more than one engine, as the capacity of the single bottle will not be sufficient for effective use on either engine in that case.

h. __Emergency Bomb Release.__ - An emergency bomb release handle is located at the pilot's left. Pulling the handle will result in immediate release of bomb door latches, and continued pulling will result in release of all bombs salvo the instant the doors reach the full open position. The bomb bay fuel tanks may also be dropped by the release handle.

i. __Pilot's Compartment Radio Controls.__

(1) __General.__

(a) All of the communications equipment may be operated to some extent from the pilot's compartment. Receiver and transmitter frequency selection of this equipment may be controlled, with the exception of the liaison equipment which must have both its transmitter and receiver frequencies set from the radio operator's position.

CAUTION: For normal operation of all communications equipment, the crystal filter selector switch should be set at "BOTH." To receive the radio range without possibility of voice interference, set the selector switch to "RANGE." To receive voice without range interference, set selector switch to "VOICE." It is impossible to receive voice when this switch is set on "RANGE."

NOTE: The headset extension cord should be plugged into the crystal filter selector control box as illustrated, and not into the interphone jackbox or the receiver control box.

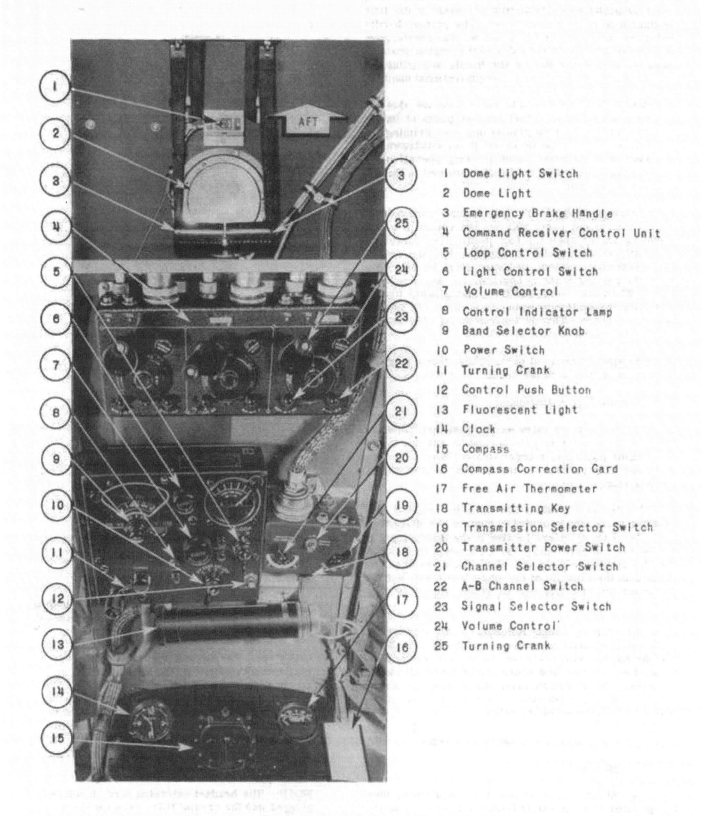

Figure 4 - Radio Controls, Ceiling, Pilot's Compartment

1 Dome Light Switch
2 Dome Light
3 Emergency Brake Handle
4 Command Receiver Control Unit
5 Loop Control Switch
6 Light Control Switch
7 Volume Control
8 Control Indicator Lamp
9 Band Selector Knob
10 Power Switch
11 Turning Crank
12 Control Push Button
13 Fluorescent Light
14 Clock
15 Compass
16 Compass Correction Card
17 Free Air Thermometer
18 Transmitting Key
19 Transmission Selector Switch
20 Transmitter Power Switch
21 Channel Selector Switch
22 A-B Channel Switch
23 Signal Selector Switch
24 Volume Control
25 Turning Crank

IMPORTANT: When the throat microphone is being used for either interphone or radio communication, it must be adjusted so that its two circular elements are held snugly against each side of the throat just above the "Adam's-apple." SPEAK SLOWLY, DISTINCTLY, AND IN A NORMAL TONE OF VOICE. Shouting will seriously distort the voice signal.

(b) A possible means of limiting noise level in all radio equipment, caused by adverse conditions such as rain, snow, ice, or sand, is to direct the radio operator to proceed as follows:

1. Place the antenna change-over switch to the fixed antenna position.

2. Release approximately 50 feet of the trailing wire antenna.

3. Ground the trailing wire antenna post directly to the airplane structure (for instance, the metal support for the transmitter tuning units).

CAUTION: Do not extend retractable rod antenna at speeds greater than 240 mph.

(2) <u>Interphone Equipment RC-36</u>. - The interphone jackbox has five selective positions marked on the face of each box, as follows:

"COMP": Reception is made through the radio compass.

"LIAISON": The pilot may transmit and receive over the liaison set.

"COMMAND": The pilot is able to transmit and receive over the command set.

"INTER": In this position the pilot may communicate with any other crew member who also has his interphone jackbox switch at the "INTER" position.

"CALL": This is an emergency position which enables any crew member to call all other members of the crew regardless of the position of their interphone jackbox switches. This position will override all other radio transmission and reception.

(3) <u>Command Set SCR-274-N</u>.

(a) <u>General</u>. - The command set is designed for short range operation and is used for communicating with nearby aircraft for tactical purposes and with ground stations for navigational and traffic control purposes.

(b) <u>Receiving</u>. - The interphone jackbox switch (figure 42-8) must first be placed in the "COMMAND" position. The receiver control box (figure 4-4) is divided into three identical sections, each section controlling the particular receiver to which it is electrically and mechanically connected. Reception of a signal of a specific frequency as indicated on the dial is accomplished by the use of the section of the receiver control box which controls the particular receiver involved. The desired receiver is turned on and off by a switch (figure 4-23) located in the left forward corner of the control box section used. This switch, in addition to having an "OFF" position, has two selective positions marked "CW" and "MCW," each of which is an <u>ON</u> position and indicates the type of signal which is to be received. The "A-B" switch (figure 4-22) should be left in the "A" position at all times and need not be turned off when the receivers are turned off

NOTE: When tuning receiver for a definite frequency, always turn dial a little to each side of the frequency calibration mark to find the point where the signal is the strongest.

(c) <u>Transmitting</u>.

1. Before transmitting, adjust radio receiver to the same frequency as the station with which you desire to talk, and listen in to be sure that the operator is not talking to someone else. If the station is transmitting, take advantage of the opportunity to more accurately set the airplane receiver on the assigned frequency, and when the other operator is finished, proceed with your transmission.

2. Throw the switch marked "OFF"-"ON" (figure 4-20) on the transmitter control box to the "ON" position. Select type of transmission desired with switch marked "TONE-CW-VOICE." (See figure 4-19) With the switch marked in the "VOICE" position, the microphone from any interphone jackbox switched to "COMMAND" position will be operative and voice will be transmitted when the push-to-talk button on the control wheel (see inset) is pressed. With the switch turned to the "CW" position, a continuous wave, or unmodulated signal, will be transmitted, and with the switch in the "TONE" position, a modulated tone signal is transmitted. Greatest effective range can be obtained on "CW." Range is most limited when operating on "VOICE."

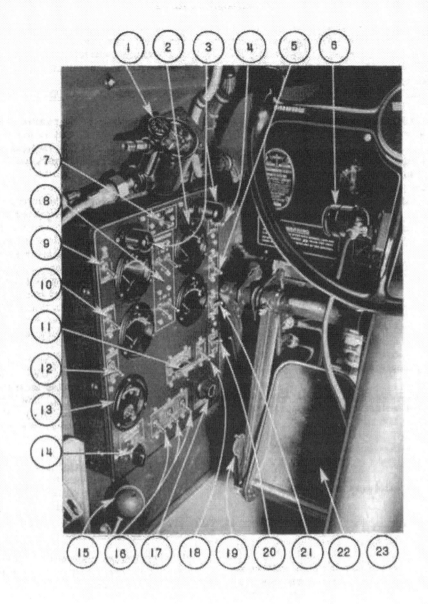

1. Oxygen Regulator
2. Ammeter
3. Panel Light
4. Signal Light Switch
5. Alarm Bell Switch
6. Ultra-Violet Spot Light
7. Panel Light Switch
8. Generator Switches
9. Passing Light Switches
10. Bomber Call Switch
11. Inverter Selector Switches
12. Phone Call Switch
13. Voltmeter
14. Voltmeter Selector Switch
15. Emergency Bomb Release
16. Master Battery Switches
17. Position Lights Control
18. Pitot Heat Switch
19. Rudder Pedal Adjustment Lever
20. Landing Gear Warning Horn Cut-off Switch
21. Hydraulic Pump Switch
22. Rudder Pedal
23. Control Column

Figure 5 - Pilot's Side Control Panel

_3_. On both the "CW" and "TONE" positions, the microphones are inoperative, and signalling by code is accomplished by a key (figure 4-18) which is located on the forward end of the transmitter control box.

NOTE: To reduce battery drain and to increase dynamotor life, the "TONE-CW-VOICE" switch should be left on "VOICE" unless continued use on "CW" or "TONE" is expected.

(4) _Radio Compass SCR-269._

(a) Set the interphone jackbox switch (figure 42-8) to the "COMP" position, if aural reception of the radio compass receiver is desired. If only visual indication is desired, the switch does not have to be set in the "COMP" position.

(b) The radio compass equipment is designed to perform the following functions:

_1_. Aural reception from the fixed antenna or from the rotatable loop. For signal reception during interference caused by precipitation static or proximity of signals, the loop will prove superior.

_2_. Aural-null directional indication of an incoming signal with the loop only in use.

_3_. Visual unidirectional indication of an incoming signal.

(c) The receiving unit is turned on or off by a switch (figure 4-10) on the face of the remote control box, which, in addition to having an "OFF" position, has three other positions: "COMP," "ANT," and "LOOP."

_1_. With the switch in the "COMP" position, both the rotatable loop and the fixed antenna are in use.

_2_. In the position marked "ANT" only the fixed antenna is in use.

_3_. With the switch turned to the "LOOP" position, only the rotatable loop is in use.

(d) If the green indicator on the face of the control box (figure 4-8) does not light, depress button marked "CONTROL" (figure 4-12) to establish control of the set at this unit. Select frequency band desired as indicated in kilocycles on the face of control box and tune by use of the crank (figure 4-11) to the desired frequency. The loop may be rotated to any position as indicated on the radio compass azimuth indicator (figure 37-3) by use of switch marked "LOOP L-R." (See figure 4-5.) This particular operation is possible only when operating on "LOOP" position of the selector switch. During periods of sever precipitation static, operate on "LOOP." For best aural reception, rotate the loop by means of the "LOOP L-R" switch (figure 4-5) until a maximum signal is obtained. Proper volume may be obtained by use of knob marked "AUDIO." (See figure 4-7.)

(5) _Marker Beacon Equipment RC-43._ - Since the operation of the marker beacon equipment is fully automatic, no manual operation is necessary. As the ship passes over a fixed point from which a marker beacon signal is being transmitted, the signal is picked up by the receiver, causing the indicator (figure 37-11) to flash on, showing the pilot that he has passed over a marker beacon. The marker beacon equipment is simultaneously turned on when the radio compass is put into operation. The position of the interphone jackbox switch does not affect the operation of the marker beacon equipment.

(6) _Liaison Set SCR-287._

(a) The liaison equipment is to be used for long range communication. Limited control is available to the pilot. The type of reception and transmission desired must be forwarded to the radio operator, who will in turn put the radio equipment in operating condition.

(b) Set the interphone jackbox switch (figure 42-8) to "LIAISON" position to receive or transmit with the liaison equipment.

(c) It is possible for all crew members to receive on this equipment, but only the pilot, copilot, and radio operator may transmit.

(7) _Radio Set SCR-535 (IFF)._ - The remote OFF-ON switch (figure 37-4) for this equipment is located on the top of the instrument panel hood. The two destroyer push button switches (figure 37-2) are located to the left of the OFF-ON switch. The destroyer switches should be used only when it is contemplated abandoning the airplane over unfriendly territory. When both destroyer push buttons are pressed simultaneously, a detonator is set off in the receiver which is located in the radio compartment. The explosion of the detonator will destroy the receiver internally. No damage should be done to either the airplane or personnel at the time of destruction of the set, but bodily contact with the receiver at the time of detonation should be avoided.

NOTE: Regeneration adjustment of the IFF set must be made on the ground prior to flight in order to insure correct operation of the equipment.

(8) _Radio Set SCR-578 (Emergency Transmitter)._ - The portable emergency transmitter, located in the aft end of the radio compartment on the right side of the airplane, is intended for use in the event of an emergency water landing. Detailed instructions will be found in paragraph 10.a. of Appendix II. (See figure 53-2.)

_j_. _Identification Lights._ - Two switches (figure 39-4) and a keying button (figure 39-7) on the central control panel permit signalling with any combination of the four lights.

k. Instrument Fluorescent Lighting System.

(1) In those airplanes that are provided with this type of lighting the pilot's and copilot's instruments are lighted by means of three ultraviolet spot lights, one mounted on each control column and one at the top of the instrument panel in the center. A fourth spot light and control switch (see inset) are provided on the ceiling to provide illumination for the clock, compass, and free air thermometer. Two types of illumination are available for the instruments with this type of spotlight: flood lighting with visible fluorescent light, and ultraviolet (invisible) activation of the luminous paint on the instrument dials. To use the fluorescent flood lighting feature, rotate the shutter to the left until the visible light falls on the instrument panel. Ultraviolet illumination is obtained by rotating the shutter in the opposite direction approximately one-quarter turn. This method of illumination eliminates objectionable window glare and reflections, since only the luminous instruments are visible.

(2) The spot lights are controlled by switches, two on the pilot's, and one on the copilot's instrument panel. Each switch is provided with a "START" position. To operate, hold the switch in the "START" position for approximately two seconds. Releasing the switch allows it to spring back to the "ON" position.

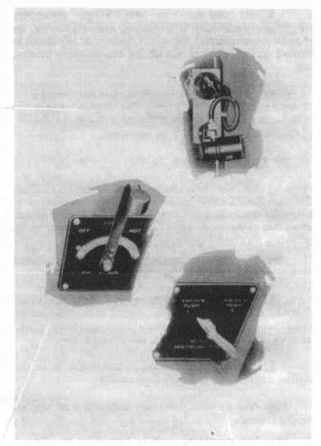

l. Alarm Bell Control.
- A toggle switch (figure 5-5) on the pilot's control panel operates three alarm bells. One bell is located in the navigator's compartment under his table, one is in the radio compartment above the radio operator's table, and one is in the tail wheel compartment inside the dorsal fin.

m. Phone Call.
- A toggle switch (figure 5-12) on the pilot's control panel operates four amber phone call signal lamps. Three of them are located adjacent to each of the alarm bells and the fourth is located in the tail gunner's compartment on the right side looking aft.

n. Bomber-Pilot Call.
- A toggle switch (figure 5-10) on the pilot's control panel operates an amber call on the bombardier's control panel, and a toggle switch on the bombardier's control panel operates an amber call lamp on the instrument panel. (See figure 37-18.)

o. Landing Gear Warning Horn Reset.
- A momentary contact toggle switch on the pilot's control panel (figure 5-20) is provided to permit the silencing of the landing gear warning horn when it is desired to continue in flight with one or more throttles closed. Operation of this switch, however, does not prevent repetition of the warning for subsequent closing of any throttle while the landing gear is up. The switch is reset when the throttles are opened.

p. Heating System Control.
- The heating and ventilating system control (see inset) is located on the side wall at the pilot's left. In the extreme forward position, hot air from the system radiator is used to ventilate the interior of the airplane. In the middle (vertical) position, all hot air is bypassed overboard and the airplane is ventilated with cold air. In the extreme aft position, no ventilating air at all enters the airplane interior. Intermediate positions of the lever between hot and cold provide ventilation with a mixture of hot and cold air, while intermediate positions between cold and off provide diminishing ventilation with cold air only.

NOTE: To provide proper heat distribution throughout the airplane, the pilot's air controls beneath the instrument panel should be approximately one-quarter open. Complete closure of the pilot's air supply is accomplished by pulling the control knob rearward.

CAUTION: Be sure the heater control is "OFF" in all ground operations.

q. Vacuum Pump Control.
- The selector valve on the side wall at the pilot's left marked "GYRO INSTRUMENTS" permits alternate use of either vacuum pump for the gyro instruments. The exhaust or pressure side of both vacuum pumps is always available for the surface de-icers, since it is not affected by the position of this selector valve. The intake or suction side is connected either to the vacuum instruments or to the surface de-icer system, depending upon the position of the selector valve.

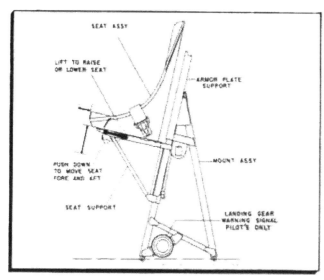

Figure 6 - Pilot's and Copilot's Seat Adjustment Diagram

r. Carburetor Air Filter Controls. - The carburetor air filter valves are operated by an electric motor installed in each assembly. Power to the motors is controlled through a four-pole double-throw relay and one double-throw toggle switch located on the instrument panel forward of the copilot. In the "CLOSED" position, induction air bypasses the filters, while in the "OPEN" position it is drawn through the filters, before entering the supercharger. Adjacent to the control switch are eight indicator lamps. When the valves are completely open, permitting only filtered air to enter the supercharger intake, four amber lamps are lighted. The four green lamps are illuminated when the control valves are fully closed, admitting only unfiltered air to the supercharger intake. Any lamp failing to illuminate should indicate that the corresponding valve has not completed its travel to the full open or full closed position.

s. De-icer Control. - The de-icer control valve controls the surface de-icer system. Operation of this valve to the "ON" position starts the distributor motor by means of an integral switch and at the same time connects pressure from both vacuum pumps, and suction from one vacuum pump to the distributor valve. In the "OFF" position the distributor valve motor is turned off, the pressure from the vacuum pumps is exhausted overboard and the suction remains connected to the distributor valve in order to keep the de-icer boots inflated.

t. Anti-icer Control. - The propeller anti-icer toggle switch on the left side wall to the rear of the pilot (figure 42-5) connects electric power to the two anti-icer pumps through two rheostats on the floor panel below it. The rheostats control the speed of the pump motors and may be used to turn the motors off if desired. Normally the rheostats should be left at a position corresponding to a predetermined rate of flow and the pump motors turned on or off by means of the toggle switch.

u. Emergency Brake System Control. - Two emergency brake handles (figure 4-3), one for each wheel, are located in the roof of the compartment just aft of the radio controls within easy reach of both pilot and copilot. A downward pull on the handle applies hydraulic pressure from an auxiliary accumulator directly to the brake mechanism.

v. Seat Adjustment. (See figure 6.)

w. Wing Flaps. - An electric motor-driven retracting mechanism operates the wing flaps through five actuating screws on each flap. The motor mechanism is located in the trailing edge of the left wing, and the torque connection for the hand crank is mounted at the forward end of the camera pit. The motor is controlled electrically by a toggle switch on the central control panel. (See figure 39-10.) The time required to lower the flaps at 147 mph is between 15 and 30 seconds. The position of the flaps at all times is shown by the indicator (figure 38-7) on the copilot's instrument panel.

WARNING: In returning the flap control switch from "DOWN" to "OFF" precautions should be taken to limit the toggle travel to the "OFF" position. If the toggle switch is allowed to snap to "OFF" a worn switch may permit the toggle to continue to the "UP" position, resulting in immediate retraction of the flaps.

x. Automatic Flight Control Equipment. - Automatic flight control equipment switches are located on the lower control panel on the front of the control stand. Switches and dials for operation of the servo units are provided. Six warning lights in front of the pilot indicate synchronization of the servo units with the control surfaces.

WARNING: Do not engage A.F.C.E. motors until all "telltale" lights are off.

y. Cowl Flaps. - A bank of four valves on the central control panel provides control for the hydraulically-operated cowl flaps. Each valve controls an actuating cylinder in the corresponding nacelle and is marked to indicate operation for opening or closing the flaps. Stops are provided within the cylinder so that the valve may be turned to "OPEN" or "CLOSE" and left there temporarily, but it is quite essential to turn the valve to "LOCKED" when the desired position of the flaps is reached, even though the full travel is required. Slight "cracking" of the control valve will result in relatively slow travel of the flaps when close adjustment is desired.

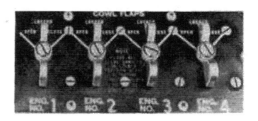

z. __Oil Dilution__. - Four momentary contact toggle switches on the side of the auxiliary panel in front of the copilot provide for oil dilution. Each switch operates a solenoid valve on the fire wall in the corresponding nacelle, which will direct fuel from the carburetor to the engine oil in line. It is obvious that in order to provide easier cold weather starting, this function must be performed AFTER an engine run immediately prior to shutting it off.

NOTE: Do not dilute oil over four minutes.

COLD WEATHER DILUTION: When operating in cold climates, the propeller control will be moved from extreme increase to extreme decrease rpm slowly several times during the period of oil dilution. This operation will permit the filling of the propeller dome with diluted oil and will prevent sluggish response of the propeller when starting the engine the next time. The supercharger controls should be operated continuously to expedite the flow of diluted oil to the regulators. With warm oil in the engine, the minimum time for operating the regulator control from the low boost to the high boost position, should be five seconds. If the oil is somewhat cooler than normal engine temperatures, this time should be extended to 15 seconds.

aa. __Fuel Boost Controls__. - The fuel boost pumps at each of the four engine fuel tanks are operated by means of four toggle switches on the central control panel. The operation of these pumps provides fuel pressure at the carburetor for starting, primes the engine fuel pumps, and also supplies pressure for priming the engines prior to starting.

ab. __Fuel Shut-off Valve Switches__. - The solenoid valves in the fuel feed lines near the booster pump at each engine tank are operated by means of four toggle switches. These valves are normally open and permit free flow of fuel except when energized by means of the toggle switches. Their purpose is to permit immediate shut-off of fuel at the tank when necessary. Failure of electrical power causes valves to "OPEN" position so that fuel will flow.

ac. __Oxygen System Filler and Relief Valves__. - The filler and relief valves for the main oxygen system are located on the left-hand side of the airplane immediately above the forward entrance hatch.

ad. __Bleeding the Brake System__. - Bleeding air from the brake system is automatically accomplished by several slow applications and complete releases of pressure through the brake metering valves. Displaced air from the brake lines will rise into the supply tank which is vented to the atmosphere.

ae. __Supercharger Regulator__. - The supercharger regulators are operated by engine oil pressure. Oil

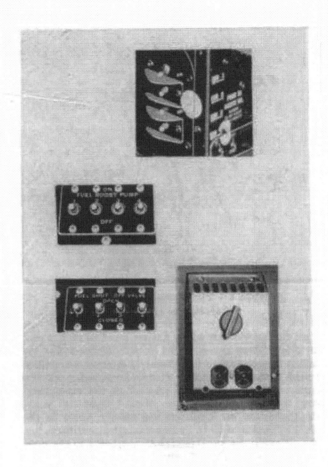

dilution serves the purpose of less delay in the proper functioning of the regulators as well as an easier engine start. During the engine oil dilution period the supercharger controls should be operated continuously to expedite the flow of diluted oil to the regulators. With warm oil in the engine the minimum time for operating the regulator control from the low boost to the high boost position should be five seconds. If the oil is somewhat cooler than normal engine temperatures, this time should be extended to 15 seconds.

af. __Fuel Indicator__. - A liquidometer indicator (figure 38-5) is located on the right-hand side of the central control panel for determining the available fuel supply. A six-position switch (figure 38-13) directly below the indicating dial, closes the electrical circuit to any one of the six regularly installed fuel tanks. Only the quantity in that tank is shown. The bomb bay tanks are not included. A warning light (figure 38-4) over the dial provides visual indication of a dangerously low fuel supply in the tank to which the indicator is turned. This light may be tested by an adjacent switch provided for that purpose.

ag. __Suit Heater Outlets__. - The heat output of each suit is controlled by a rheostat in the receptacle box.

ah. __Use of Oxygen__.

Use Oxygen Intelligently

DO...

USE Oxygen Above 10,000 Feet on All Flights

USE Oxygen from the Ground Up, at Night, or on Rapid Ascents to High Altitude

Breathe Normally

ADJUST Your Mask Carefully and Eliminate Leaks Before Take-Off

BE Thoroughly Conversant with Your Oxygen Equipment and Reasons for its Use

REPORT Faulty Function of Oxygen Equipment Promptly and Insure Correction

CHECK Your Oxygen Equipment Frequently During Flight

DO NOT...

DO NOT Fail to Check All Oxygen Equipment Before Take-Off

DO NOT Fail to Insure Full Cylinder Pressure and an Adequate Supply Oxygen for Your Mission

DO NOT Fail to Use Your Own Fitted Mask and Necessary Connecting Tubing

DO NOT Leave Your Walk-Around Bail-Out Oxygen Bottles in Your Locker. You May Need Them

DO NOT Waste Your Oxygen Supply by Excessive and Needlessly High Flows

DO NOT Take Liberties at High Altitude by Walking About the Aircraft Without Portable Oxygen Bottles, or by Not Turning on the Oxygen Supply in Time

RESTRICTED

ai. Load Adjuster.

(1) *Application of Load Adjuster*. - A load adjuster and carrying case for the model B-17F airplane will be found located on a mounting hook adjacent to the data case. Pick up the instrument and ascertain that the serial number for airplane being loaded is identical with the serial number inscribed on the carrying case identification card (view A).

CAUTION: The airplane model designation stamped on every load adjuster indicates that the instrument may be used for balance calculations on any AAF airplane of that particular model. However, the index figure entered on the carrying case identification card is correct only for the specific airplane serial number printed directly above, and represents the balance moment of only that one individual basic airplane.

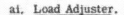

- 12 -

RESTRICTED

RESTRICTED   T. O. No. 01-20EF-1

(2) <u>Operating Instructions</u>.

(a) The following sample loading problem is itemized in detail, and complete instructions, with supporting illustrations, are published to furnish the Service with complete instructions on loading aircraft above and beyond the basic airplane (including personal items and all items commonly referred to as "disposable"). All items are to be taken into consideration for each and every loading problem, and their balance moments must be added with the load adjuster on the compartment scales where they are located.

<u>1</u>. Given:

Basic Airplane		34,500
Oil (150 U.S. gal)		1,125
Gasoline (1732 U.S. gal)		10,392
Nose Compartment		500
Bombardier	200	
Navigator	200	
Ammunition (800 rd .30 cal)	52	
Handbook and Map Data	48	
Pilot's Compartment		450
Pilot and Copilot (200 lb ea)	400	
Baggage (Below Compartment)	50	
Upper Turret Compartment		530
Aerial Engineer	200	
Ammunition (1000 rd .50 cal)	330	
Bombs		5,000
Radio Compartment		700
Radio Operator	200	
Ammunition (2 boxes stowed)	350	
Special Equipment	150	
Ball Turret		530
Gunner	200	
Ammunition	330	
Side Gun Compartment		530
Gunner	200	
Ammunition (1000 rd .50 cal)	330	
Door Compartment (Special Equipment)		330
Tail Gun Compartment		400
Gunner	200	
Ammunition (610 rd .50 cal)	200	
GROSS WEIGHT		54,987

Revised 4-1-43   - 13 -   RESTRICTED

2. To find: If the load distribution brings the airplane balance within permissible cg limits as indicated on the load adjuster "Loading Range" scale.

VIEW B   a. Set indicator hairline on basic airplane index 22.4 (obtained from identification card on load adjuster carrying case) and move slide to zero mark on the "OIL" scale as illustrated in view B.

VIEW C   Move indicator until the hairline is over 150 on the "U.S. GALS" scale. This adds the balance moment of 150 gallons of oil as loaded in the airplane's oil tanks, and moves the index to 14.9 as shown in view C.

VIEW D   b. Set slide to the zero mark on the "GASOLINE" scale as shown in view D.

VIEW E   Move indicator until the hairline is over 1732 on the "U.S. GALS" scale. This adds the balance moment of 1732 gallons of gasoline as loaded in the airplane's tanks and moves the index to 28.5 as illustrated in view E.

VIEW F   c. Set slide to the compartment zero line ("STATION LOADS") as illustrated in view F.

VIEW G   Move indicator until the hairline is over 500 on the "NOSE" compartment scale. This adds the balance moment of two men, ammunition and data as loaded in the nose compartment, and moves the index to 18.6 as shown in view G.

VIEW H   d. Set slide to the compartment zero line as illustrated in view H.

VIEW I   Move indicator until the hairline is over 450 on the "PILOTS" scale. This adds the balance moment of the two pilots and baggage below the pilot's compartment and moves the index to 14.2 as illustrated in view I.

VIEW J   e. Set slide to the compartment zero line as illustrated in view J.

VIEW K Move indicator until the hairline is over 530 on the "TOP TURRET" scale. This adds the balance moment of one man and 1000 rounds of .50 caliber ammunition in the top turret and moves the index to 10.8 as illustrated in view K.

VIEW L f. Set the slide to the compartment zero line as shown in view L.

VIEW M Move indicator until the hairline is over 5000 on the "BOMB BAY" scale. This adds the balance moment of the load in the bomb bay and moves the index to 8.5 as shown in view M.

VIEW N g. Set slide to the compartment zero line as illustrated in view N.

VIEW O Move indicator until the hairline is over 700 on the "RADIO" scale. This adds the balance moment of one man, one box of .50 caliber ammunition stowed, and 150 pounds of special equipment in the radio compartment and moves the index to 15.5 as shown in view O.

VIEW P h. Set slide to the compartment zero line as illustrated in view P.

VIEW Q Move indicator until the hairline is over 530 on the "BALL TURRET" scale. This adds the balance moment of one man and 1000 rounds of .50 caliber ammunition in the ball turret and moves the index to 24.8 as illustrated in view Q.

VIEW R i. Set slide to the compartment zero line as shown in view R.

VIEW S Move indicator until the hairline is over 530 on the "SIDE GUN" scale. This adds the balance moment of the gunner and 1000 rounds of .50 caliber ammunition and moves the index to 38.6 as illustrated in view S.

VIEW T j. Set slide to the compartment zero line as shown in view T.

VIEW U Move indicator until hairline is over 330 on the "DOOR" scale. This adds the balance moment of the special equipment loaded opposite the main entrance door and moves the airplane index to 50.2 as illustrated in view U.

VIEW V k. Set slide to the compartment zero lines as shown in view V.

VIEW W Move indicator until hairline is over 400 on the "TAIL" scale. This adds the balance moment of one man and 610 rounds of .50 caliber ammunition in the tail gun compartment, and completes the calculation of the balance moments of all items as initially loaded in the airplane. It has moved the airplane index to 73.1 as shown in view W.

l. Balance Correction.

(1) Adding the weights of all items loaded (paragraph 1.c. of this section), shows the gross load of the airplane well within allowable limits, and, as far as weight alone is concerned, the airplane may be flown. However, the load adjuster indicator hairline is located in the red portion of the loading range which ABSOLUTELY PROHIBITS any attempt to fly the airplane because of a dangerous tail heavy condition.

(2) This "out of balance" condition may be corrected by shifting some of the load or a member of the crew from an aft position to a forward position in the airplane, the amount of change required being predetermined by a "trial shift" of the load on the load adjuster. It will be noted that it is prohibited to take off with any crew member in the ball turret. In this sample case, shifting the ball turret gunner from his ball turret position to the nose compartment will bring the airplane balance within allowable cg limits. This is determined as outlined in the next paragraph.

- 18 -

RESTRICTED

VIEW X (3) With the indicator hairline remaining on the last index (73.1), move the slide until "BALL TURRET" of the "CREW CHANGE - ONE MAN - 200 POUNDS" scale is under the indicator hairline as illustrated in view X.

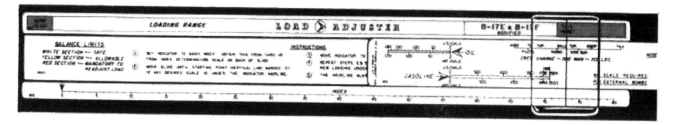

VIEW Y Move the indicator until its hairline is over "NOSE" of the crew change scale. This changes the balance moment of one man (200 pounds) from the ball turret to the nose compartment and moves the index to 65.6 as shown in view Y.

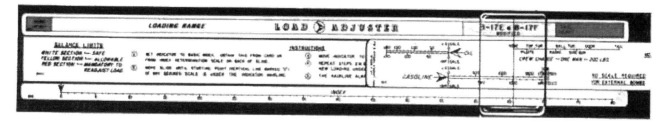

m. The airplane is now within permissible cg limits although the indicator hairline is still in the yellow portion of the "LOADING RANGE" scale. The pilot will notice a slight tail heaviness when the indicator hairline is in this yellow portion, but it may be corrected with tab adjustment, and is satisfactory for flight. More perfect balance could still be obtained by moving an additional member of the crew forward.

CAUTION: Do NOT shift or dispose of any load without first predetermining (by use of the load adjuster) that the balance will remain within limits after the change is made.

aj. Hydraulic Pump Control. - In most airplanes, power is supplied to the automatic pressure switch REGARDLESS OF THE POSITION OF THE HYDRAULIC PUMP SWITCH on the pilot's control panel. In case the automatic pressure switch fails to function, pressure may be restored by holding the hydraulic pump switch in the "MANUAL" position. In other airplanes the hydraulic pump motor "ON-OFF" switch on the pilot's control panel must be in the "ON" position to maintain hydraulic pressure automatically.

WARNING: To prevent loss of fluid should leakage occur in the hydraulic system, the pump motor must be stopped. The reccommended procedure is the removal of the 15 ampere hydraulic pump switch fuse in the station 4 fuse panel, or the disconnection of the electrical receptacle at the pressure switch.

B-17F Flying Fortress

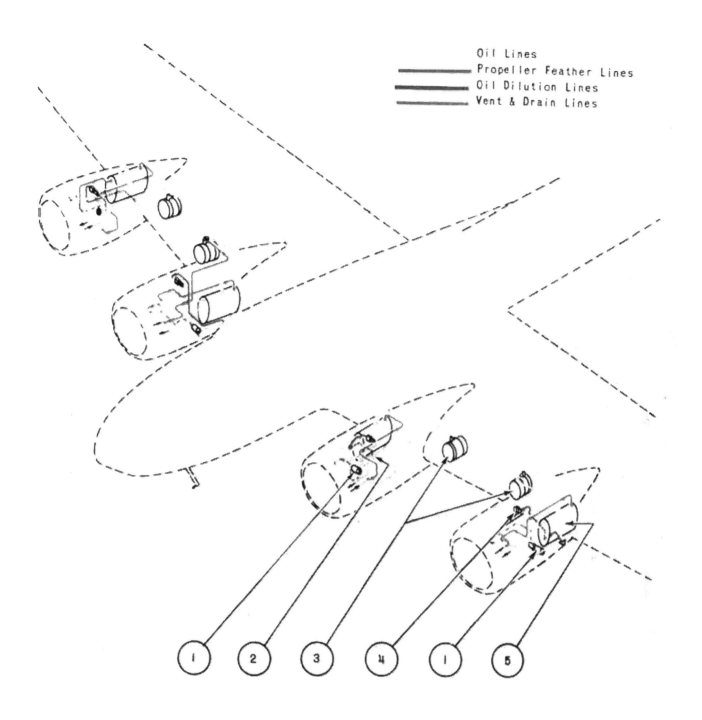

1. Propeller Feathering Pump
2. Drain Valve
3. Oil Temperature Regulator
4. Oil Dilution Valve
5. Oil Tank

Figure 8 - Oil System Diagram

RESTRICTED

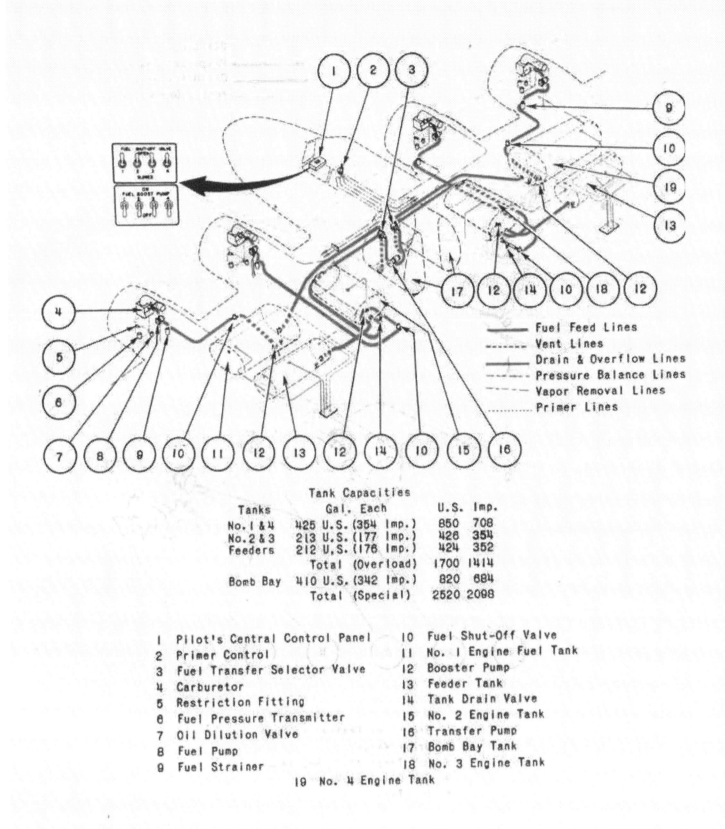

Tank Capacities

Tanks	Gal. Each	U.S.	Imp.
No. 1 & 4	425 U.S. (354 Imp.)	850	708
No. 2 & 3	213 U.S. (177 Imp.)	426	354
Feeders	212 U.S. (176 Imp.)	424	352
	Total (Overload)	1700	1414
Bomb Bay	410 U.S. (342 Imp.)	820	684
	Total (Special)	2520	2098

1. Pilot's Central Control Panel
2. Primer Control
3. Fuel Transfer Selector Valve
4. Carburetor
5. Restriction Fitting
6. Fuel Pressure Transmitter
7. Oil Dilution Valve
8. Fuel Pump
9. Fuel Strainer
10. Fuel Shut-Off Valve
11. No. 1 Engine Fuel Tank
12. Booster Pump
13. Feeder Tank
14. Tank Drain Valve
15. No. 2 Engine Tank
16. Transfer Pump
17. Bomb Bay Tank
18. No. 3 Engine Tank
19. No. 4 Engine Tank

Figure 9 - Fuel System Diagram

RESTRICTED T. O. No. 01-20EF-1

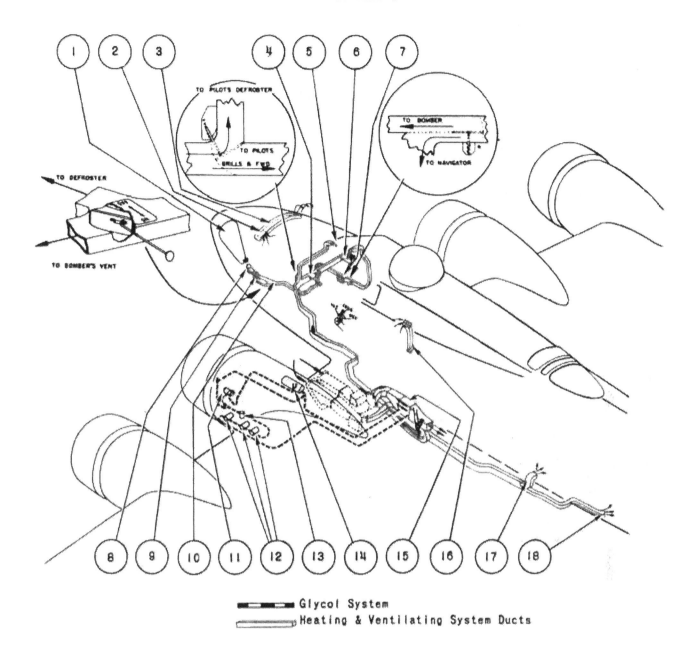

- Glycol System
- Heating & Ventilating System Ducts

1 Bombardier's Window	10 Bombardier's Air & Defroster Control
2 Bombardier's Compartment Grill	11 Pumps
3 Bombardier's Compartment Vent	12 Heaters
4 Pilot's Air Control	13 Filter
5 Pilot's Defroster and Control	14 Tank
6 Copilot's Air Control	15 Radiator
7 Navigator's Air Control and Outlet	16 Pilot's Ventilating Duct
8 Bombardier's Window Defroster	17 Radio Compartment Air Outlet & Controls
9 Bombardier's Air Duct	18 Side Gun Compartment Heating Duct

Figure 10 - Heating and Ventilating System Diagram

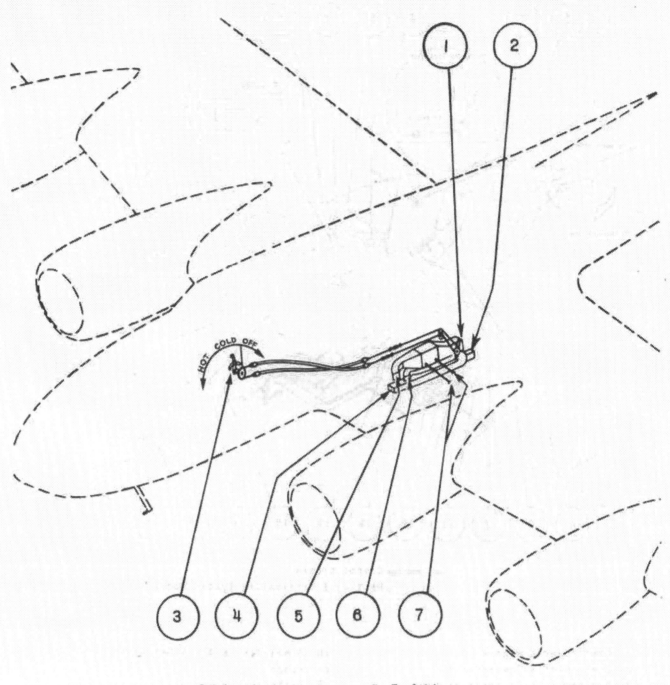

1 Overboard
2 To Aft Compartment
3 Cabin Heat Control Handle
4 To Cabin
5 Air Inlet
6 Radiator
7 To Glycol System

Figure 11 - Heating Control System Diagram

1 Glycol Tank
2 Intercooler Air Inlet
3 Carburetor Air Inlet
4 Relief Valve
5 Pump
6 Filter
7 Boilers

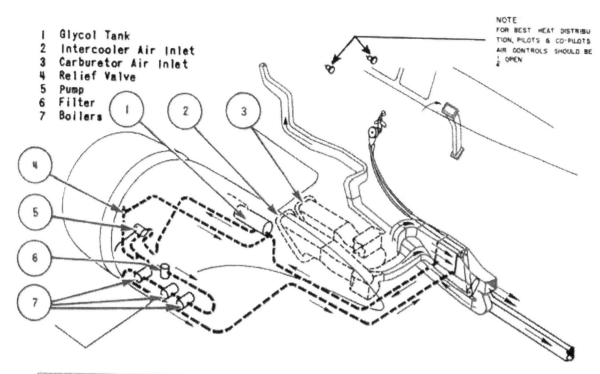

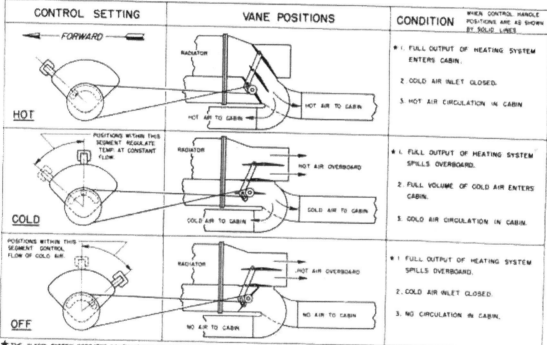

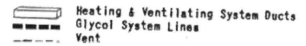

Figure 12 - Heating System Operation Diagram

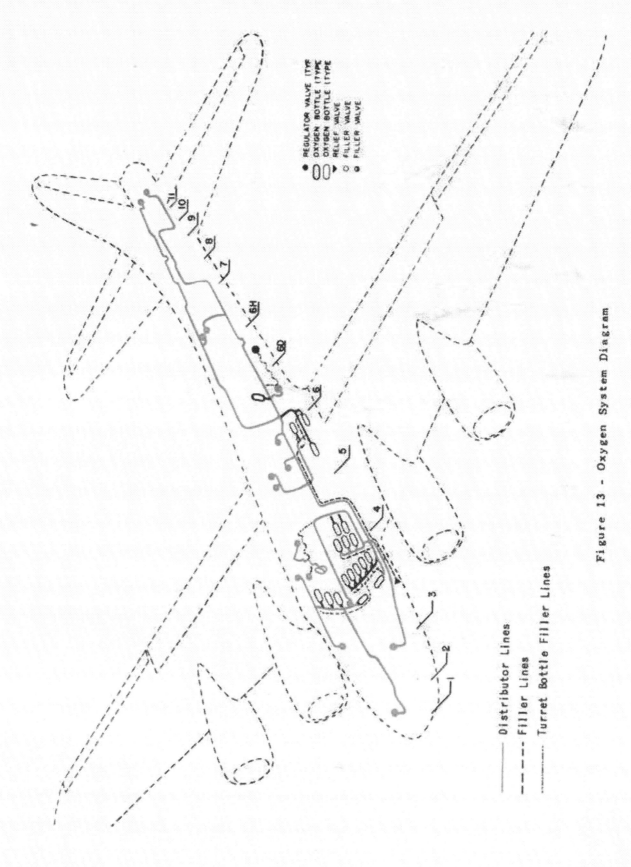

Figure 13 - Oxygen System Diagram

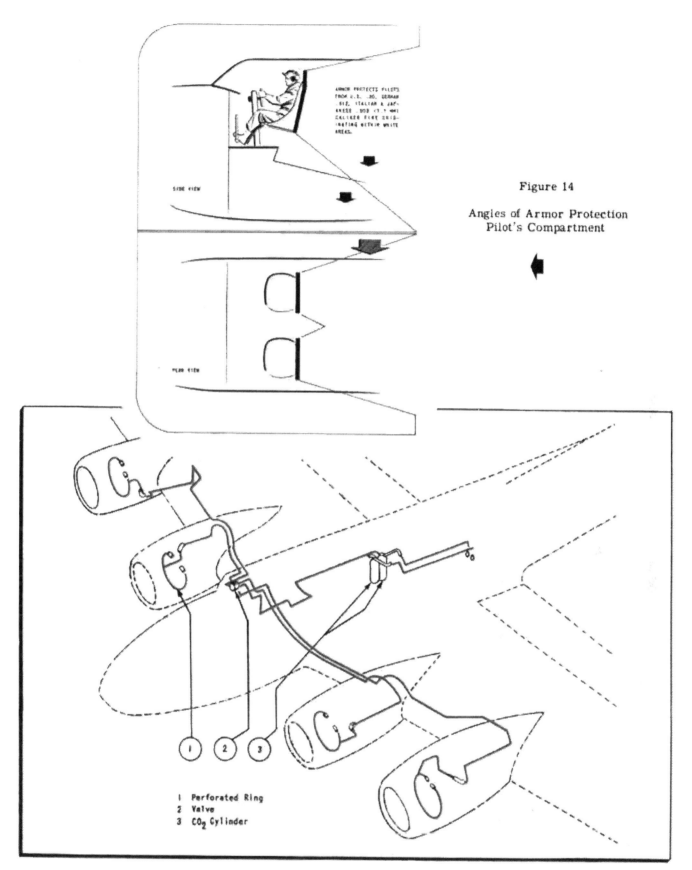

Figure 14

Angles of Armor Protection Pilot's Compartment

1 Perforated Ring
2 Valve
3 CO_2 Cylinder

Figure 15 - Engine Fire Extinguisher System Diagram

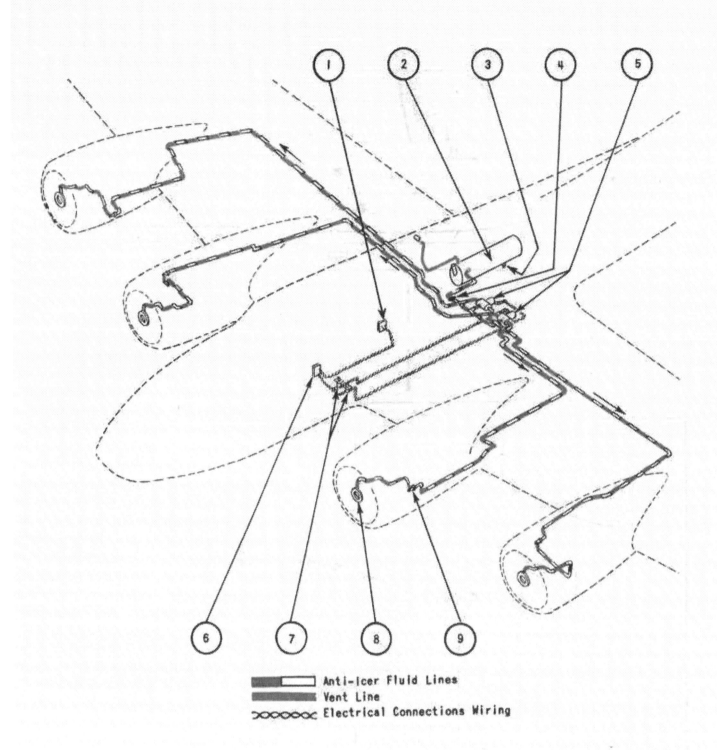

	Anti-Icer Fluid Lines
	Vent Line
⋈⋈⋈⋈	Electrical Connections Wiring

1 Fuse Panel
2 Fluid Tank
3 Drain
4 Shut-Off Valve
5 Pump
6 Pilot's Switch Panel
7 Pilot's Rheostat Controls
8 Slinger Ring
9 Valve

Figure 16 - Propeller Anti-Icer System Diagram

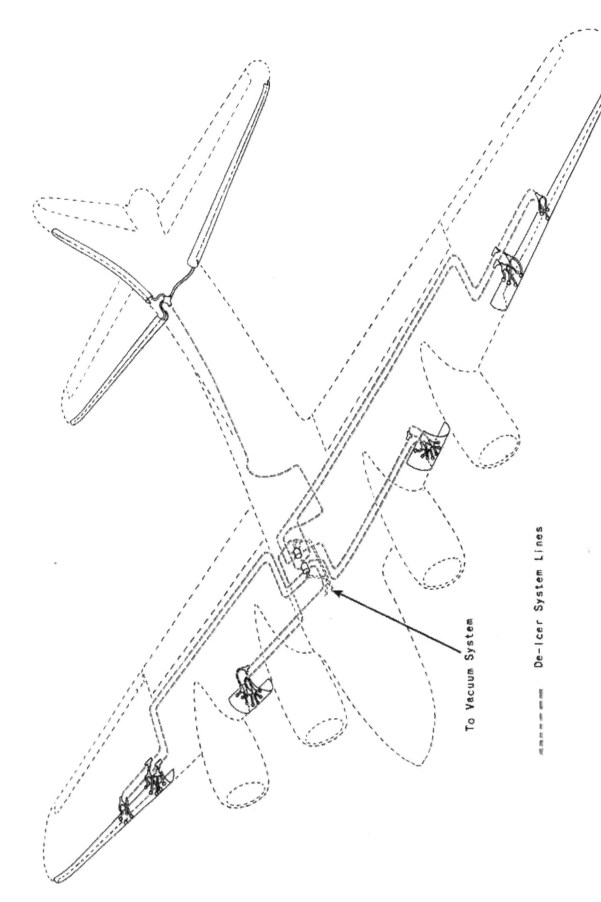

Figure 17 - Surface De-Icer System Diagram

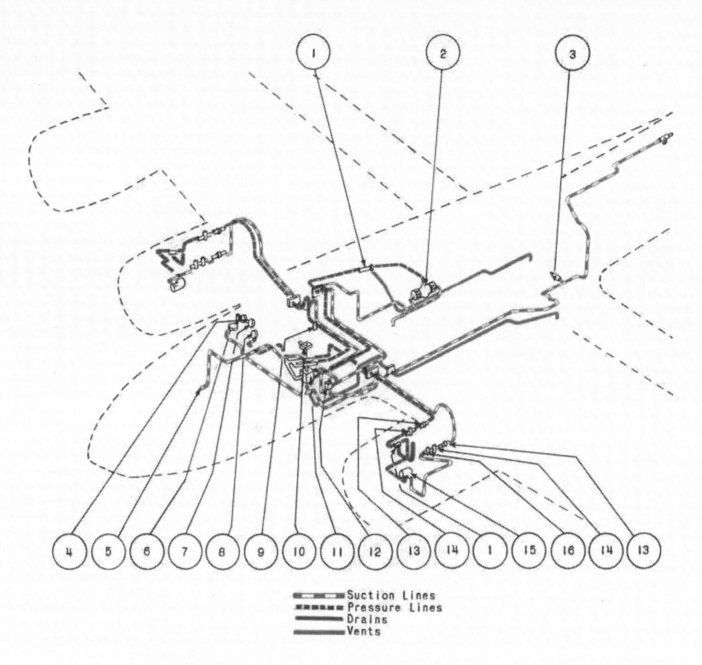

1. Oil Separator
2. Rotary Distributing Valve
3. Shut-off Valve
4. Flight Indicator
5. To Driftmeter
6. Manifold
7. Suction Gage
8. De-icer Pressure Gage
9. Test Plug
10. De-icer Control Valve Switch
11. De-icer Control Valve
12. Selector Valve
13. Check Valve
14. Pressure Relief Valve
15. Vacuum Pump
16. Suction Relief Valve

Figure 18 - Vacuum and De-Icer System Diagram

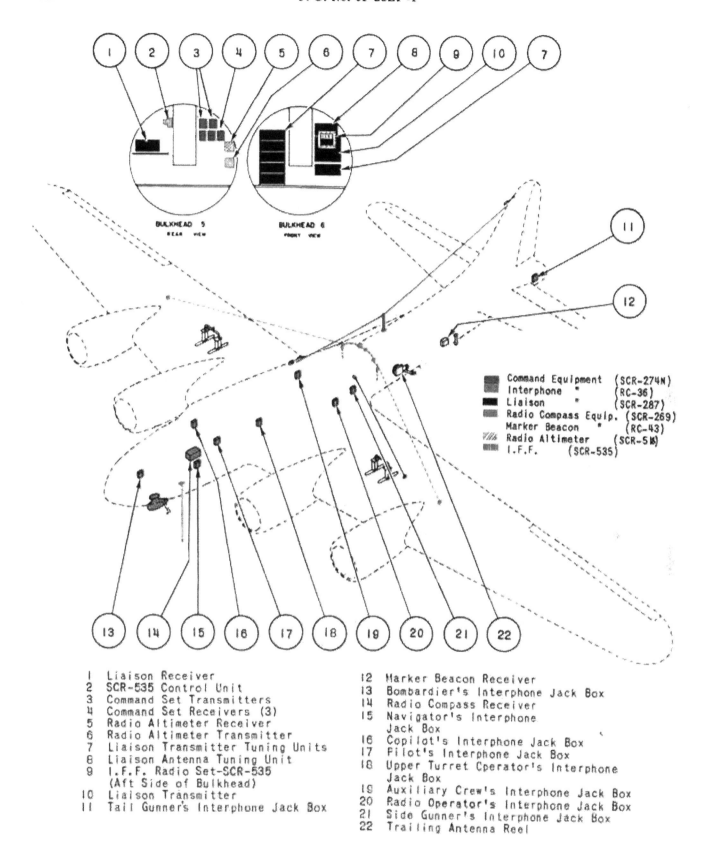

1. Liaison Receiver
2. SCR-535 Control Unit
3. Command Set Transmitters
4. Command Set Receivers (3)
5. Radio Altimeter Receiver
6. Radio Altimeter Transmitter
7. Liaison Transmitter Tuning Units
8. Liaison Antenna Tuning Unit
9. I.F.F. Radio Set-SCR-535 (Aft Side of Bulkhead)
10. Liaison Transmitter
11. Tail Gunner's Interphone Jack Box
12. Marker Beacon Receiver
13. Bombardier's Interphone Jack Box
14. Radio Compass Receiver
15. Navigator's Interphone Jack Box
16. Copilot's Interphone Jack Box
17. Pilot's Interphone Jack Box
18. Upper Turret Operator's Interphone Jack Box
19. Auxiliary Crew's Interphone Jack Box
20. Radio Operator's Interphone Jack Box
21. Side Gunner's Interphone Jack Box
22. Trailing Antenna Reel

Figure 19 - Communication Equipment Diagram

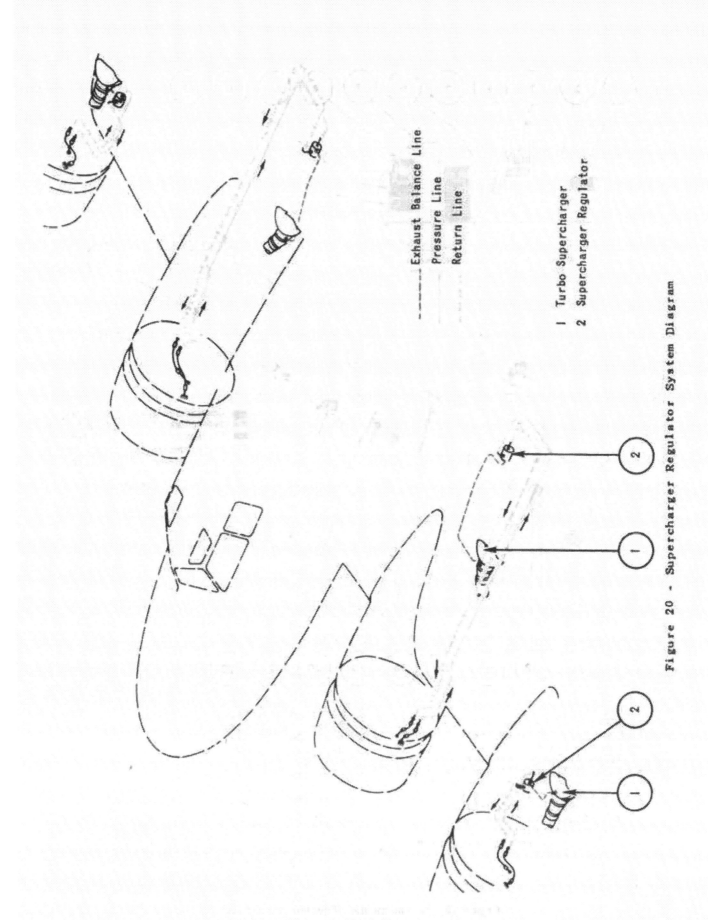

Figure 20 - Supercharger Regulator System Diagram

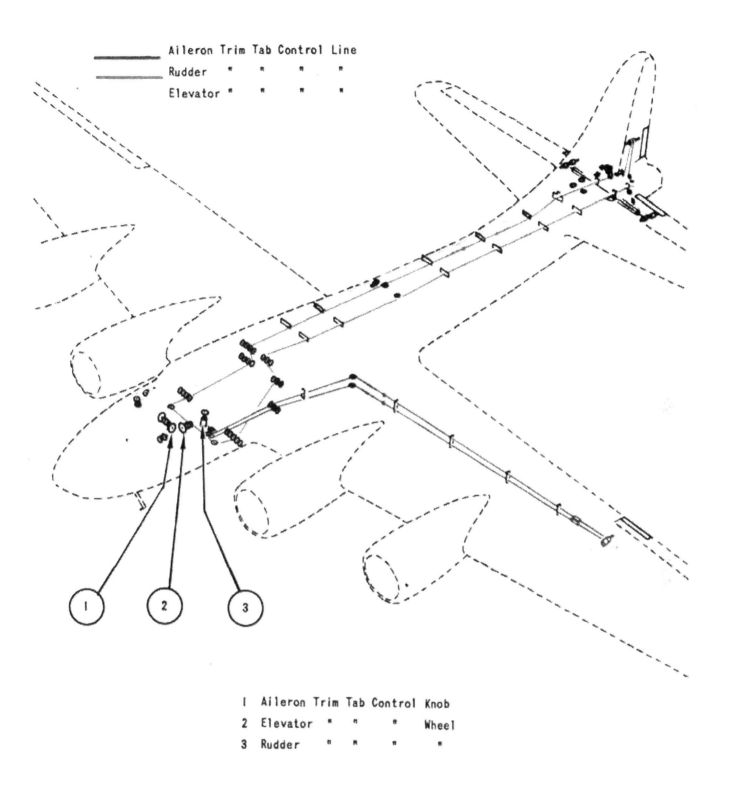

Figure 21 - Trim Tab Control System Diagram

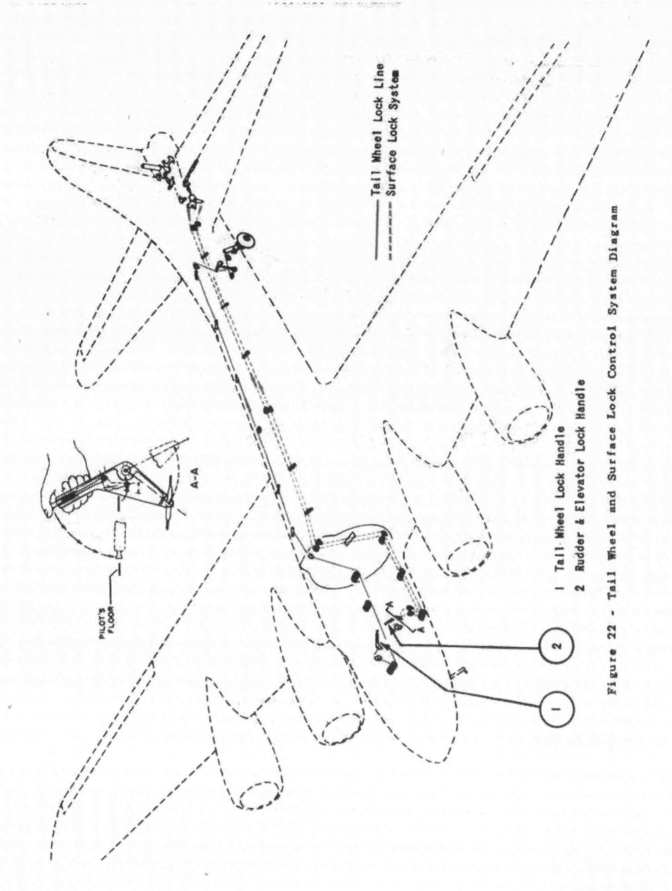

Figure 22 - Tail Wheel and Surface Lock Control System Diagram

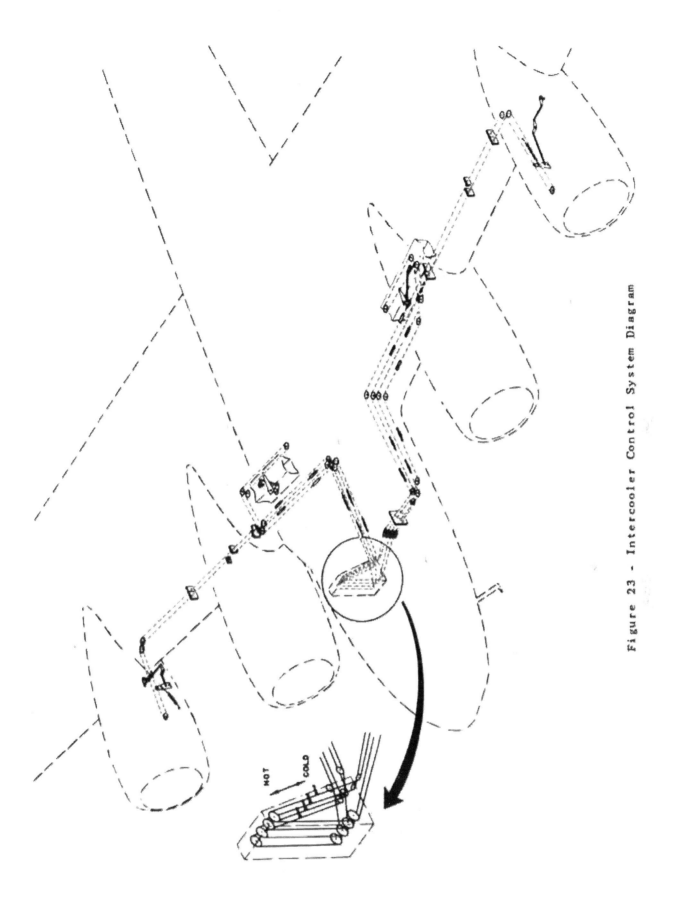

Figure 23 - Intercooler Control System Diagram

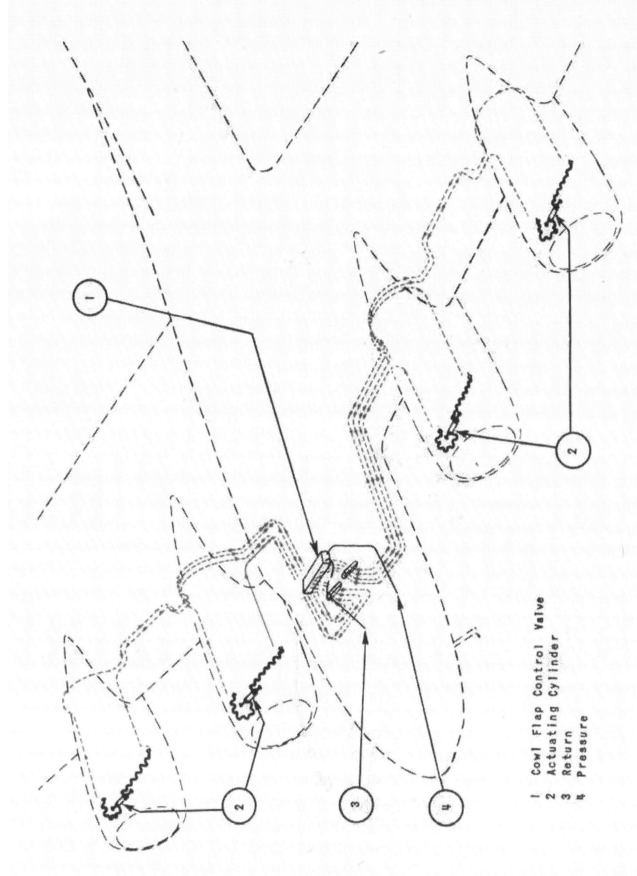

Figure 24 - Cowl Flap Control System Diagram

1 Cowl Flap Control Valve
2 Actuating Cylinder
3 Return
4 Pressure

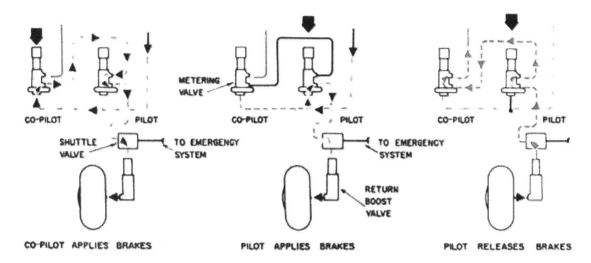

NORMAL BRAKE OPERATION

IN ORDER TO SIMPLIFY THE BRAKE SYSTEM, IT HAS BEEN DESIGNED IN SUCH MANNER THAT PRESSURE TO THE BRAKES IS APPLIED ONLY THRU THE PILOT'S METERING VALVES, AND OIL IS RETURNED TO THE TANK THRU THE CO-PILOT'S METERING VALVES. THUS WHEN THE CO-PILOT DESIRES TO OPERATE THE BRAKES, PRESSURE FROM HIS VALVE PASSES THRU THE PILOT'S BEFORE REACHING THE BRAKES. CONVERSELY, IF THE PILOT HAS APPLIED PRESSURE TO THE BRAKES, OIL WILL FLOW BACK TO THE TANK THRU THE CO-PILOT'S VALVES WHEN THE BRAKES ARE RELEASED.

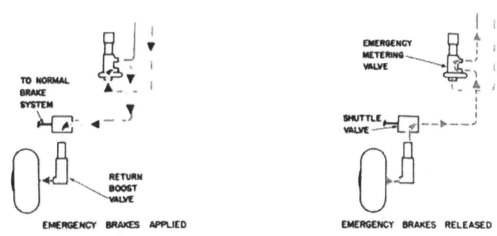

EMERGENCY BRAKE OPERATION

PRESSURE FOR THE EMERGENCY BRAKES IS SUPPLIED FROM THE AUXILIARY ACCUMULATOR AND PASSES THRU THE EMERGENCY METERING VALVES TO THE SHUTTLE VALVES. THE SHUTTLE VALVES PERMIT PRESSURE TO BE APPLIED TO THE BRAKES FROM THE EMERGENCY HYDRAULIC LINE INSTEAD OF THE NORMAL HYDRAULIC LINE. WHEN BRAKES ARE RELEASED, OIL RETURNS THRU THE EMERGENCY BRAKE LINES AND THE METERING VALVES TO THE SUPPLY TANK.

- - Pressure Line
——— Return Line

Figure 25 - Brake Operation Diagram

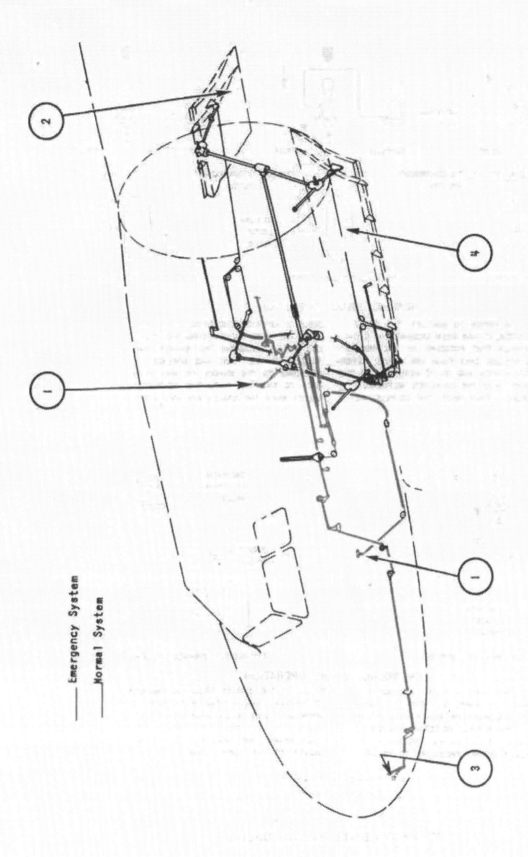

1 Emergency Bomb Release Handle
2 Bomb Bay Door Open
3 Bombardier's Bomb Release Control
4 Bomb Bay Door Closed

Figure 26 - Bomb Control System Diagram

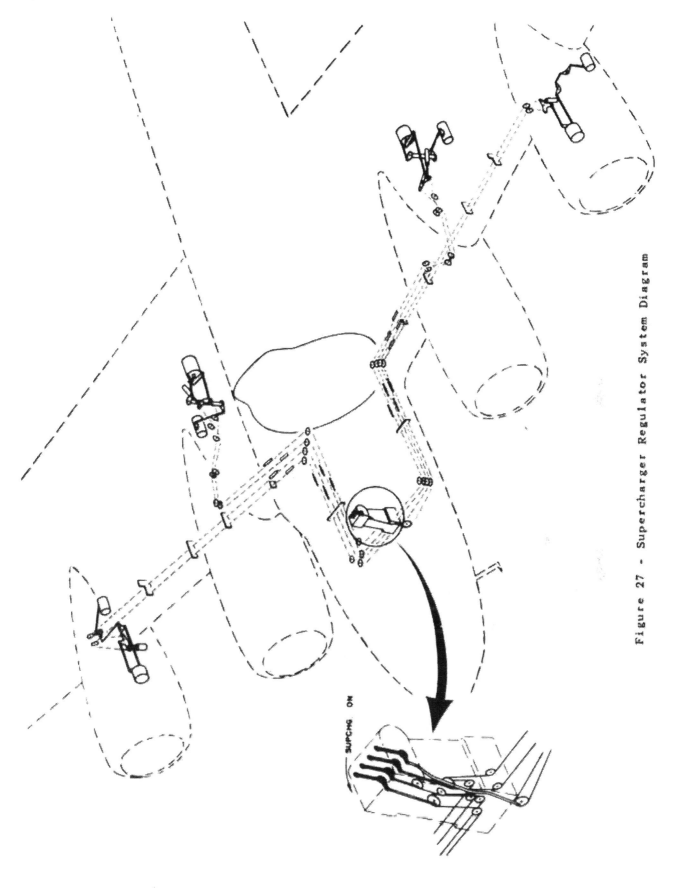

Figure 27 - Supercharger Regulator System Diagram

- 39 -

RESTRICTED

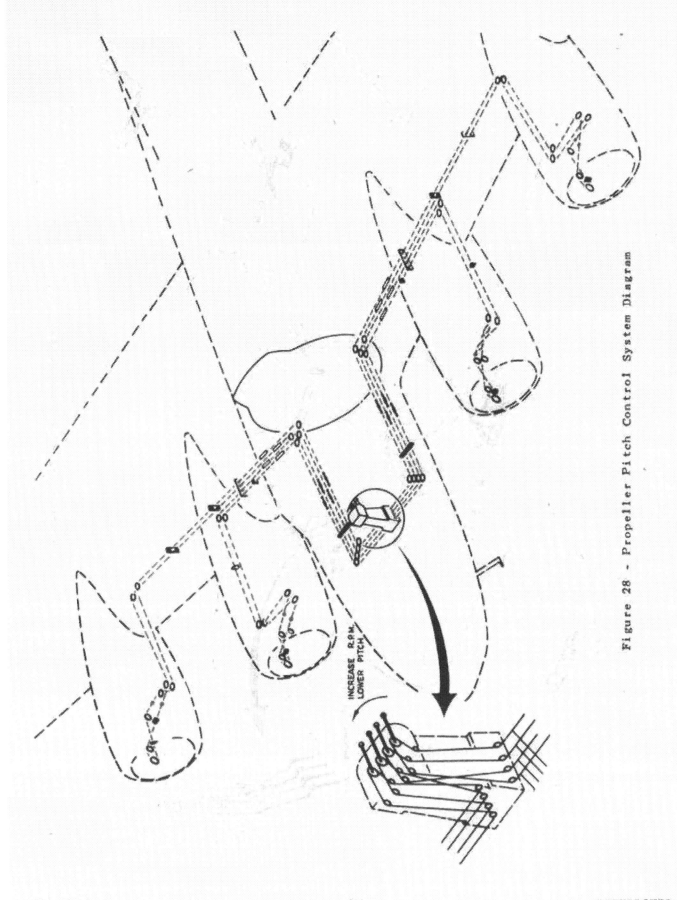

Figure 28 - Propeller Pitch Control System Diagram

- 40 -

RESTRICTED

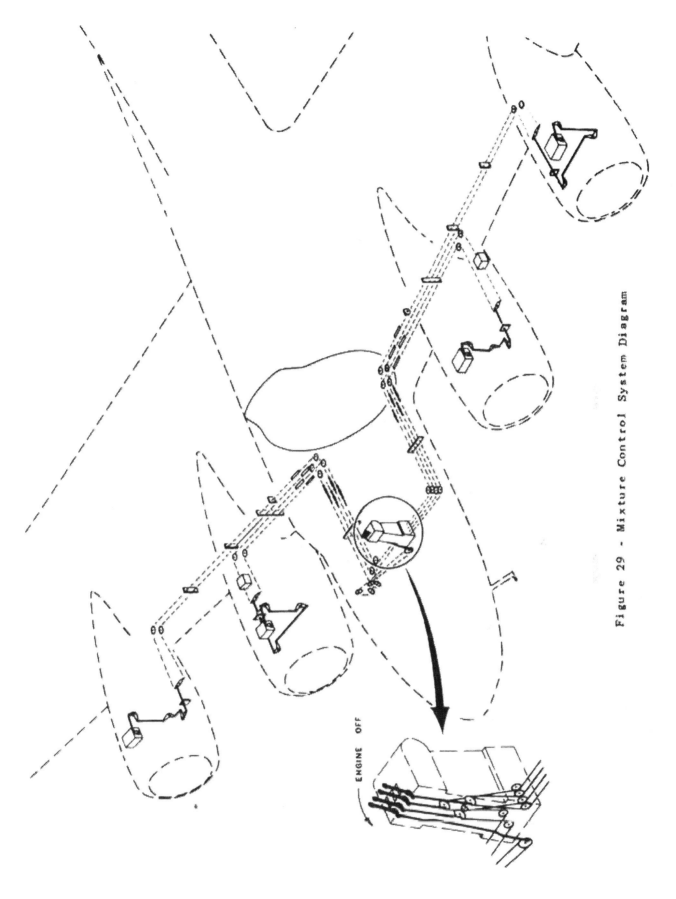

Figure 29 - Mixture Control System Diagram

RESTRICTED

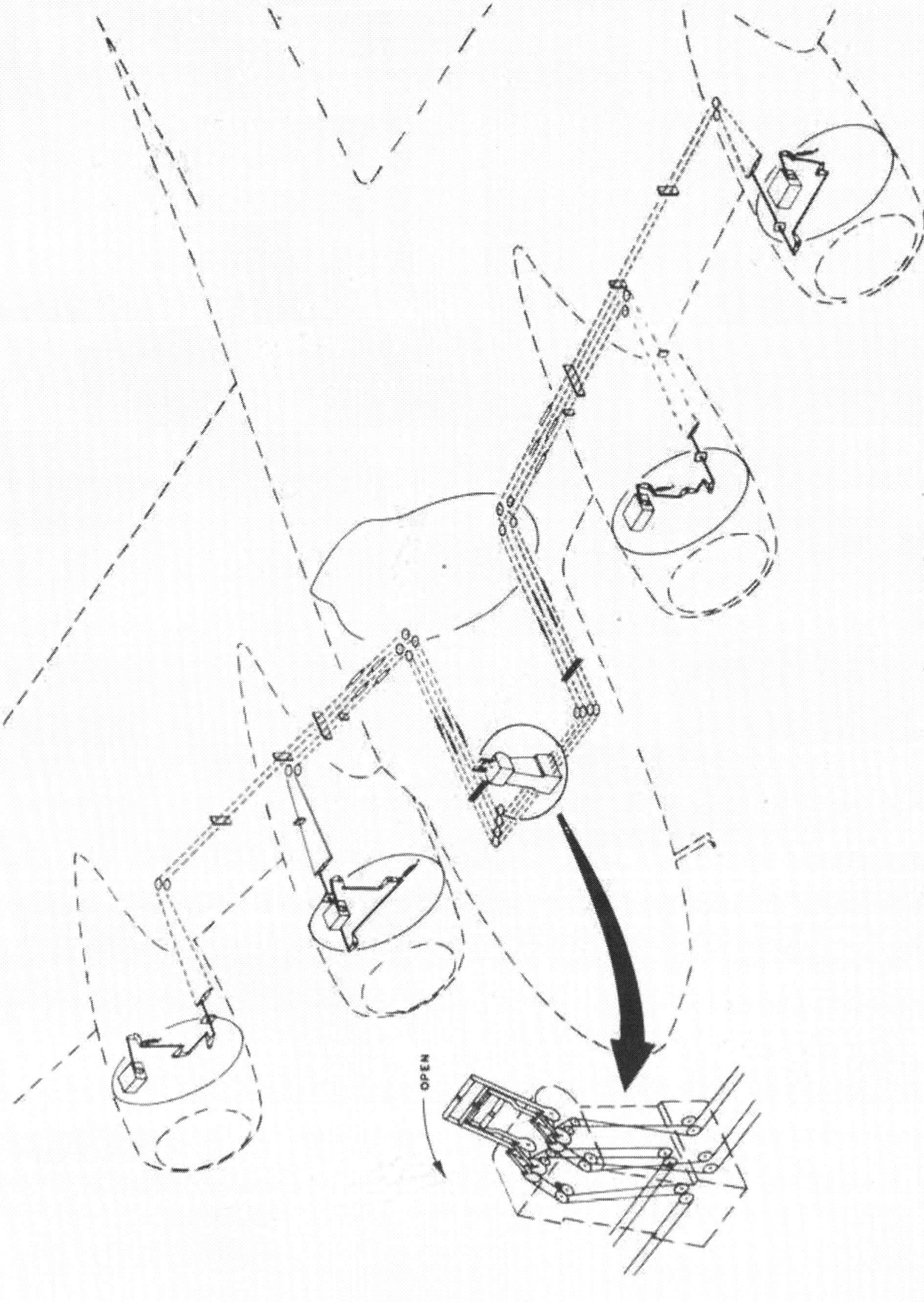

Figure 30 - Throttle Control System Diagram

- 42 -

RESTRICTED

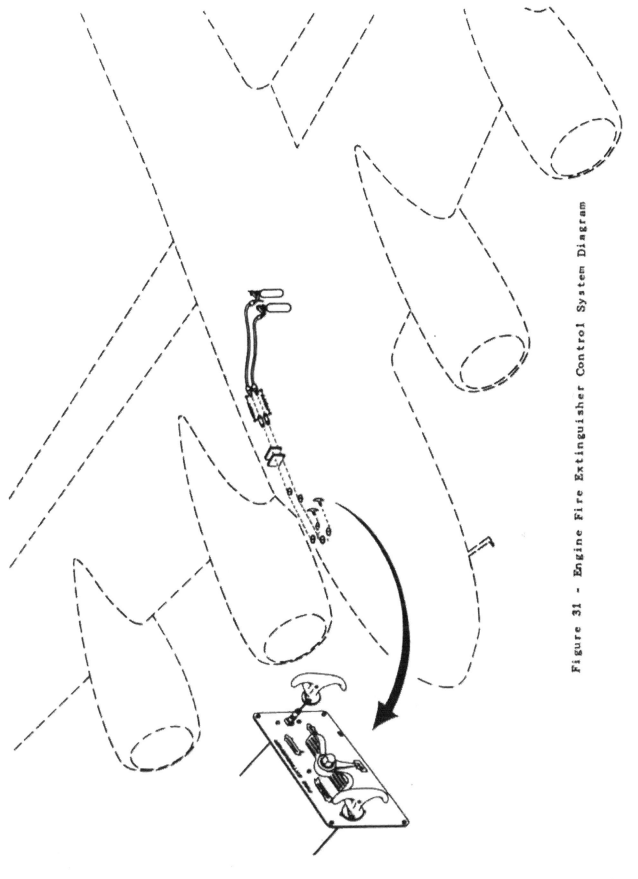

Figure 31 - Engine Fire Extinguisher Control System Diagram

RESTRICTED

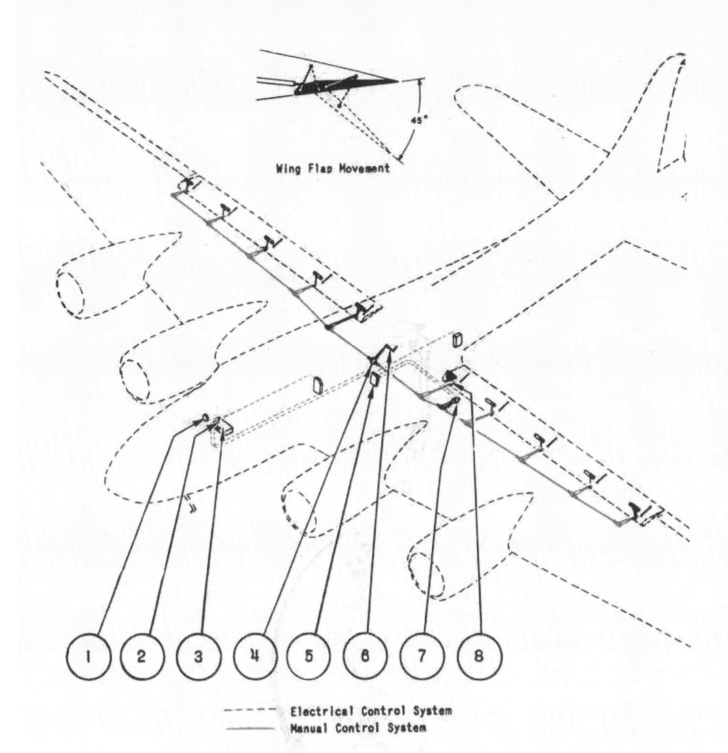

Figure 32 - Wing Flap Control System Diagram

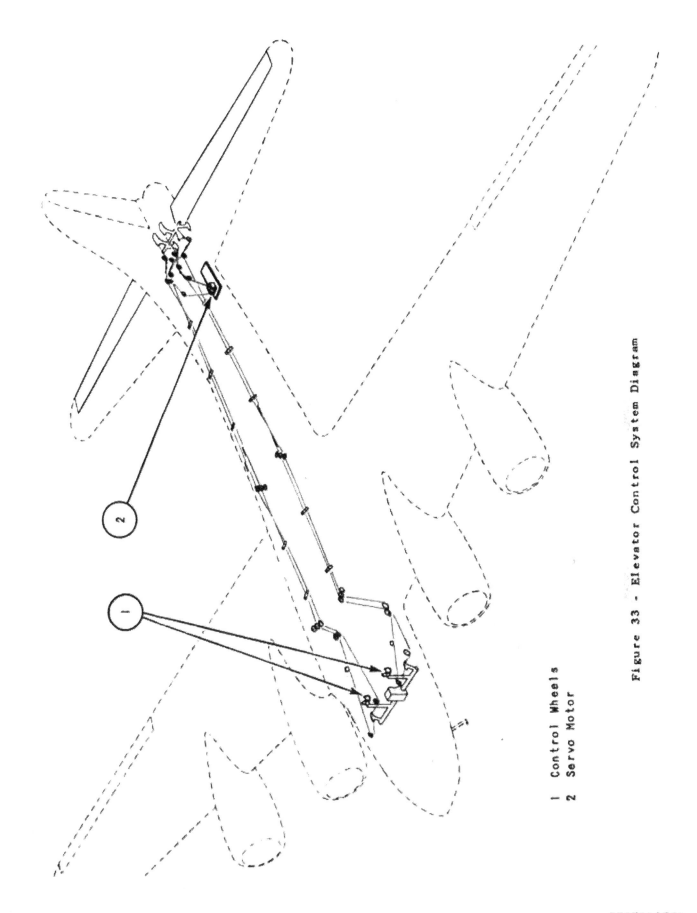

1 Control Wheels
2 Servo Motor

Figure 33 - Elevator Control System Diagram

RESTRICTED

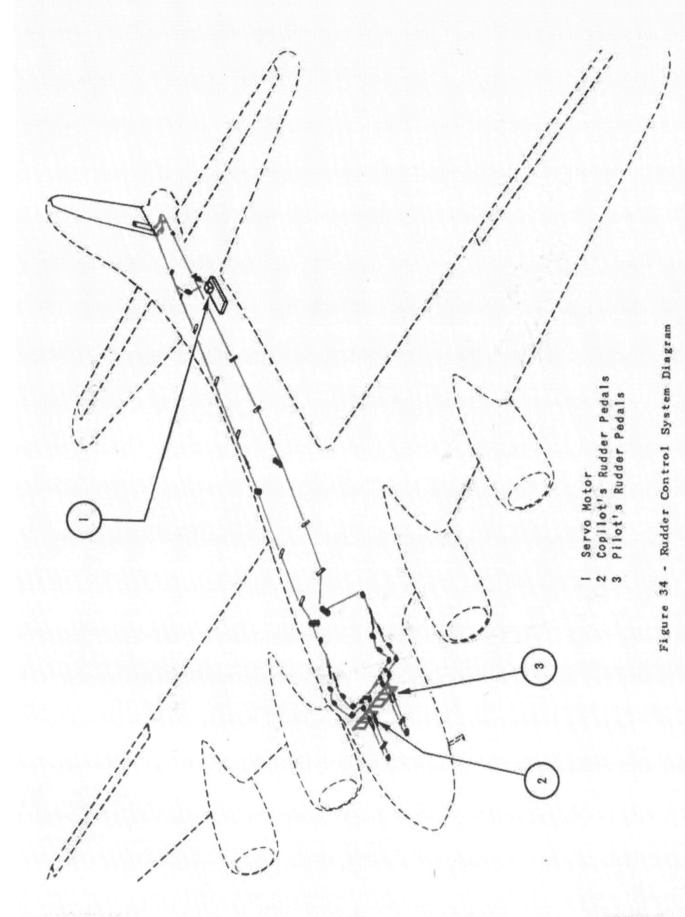

Figure 34 - Rudder Control System Diagram

1 Servo Motor
2 Copilot's Rudder Pedals
3 Pilot's Rudder Pedals

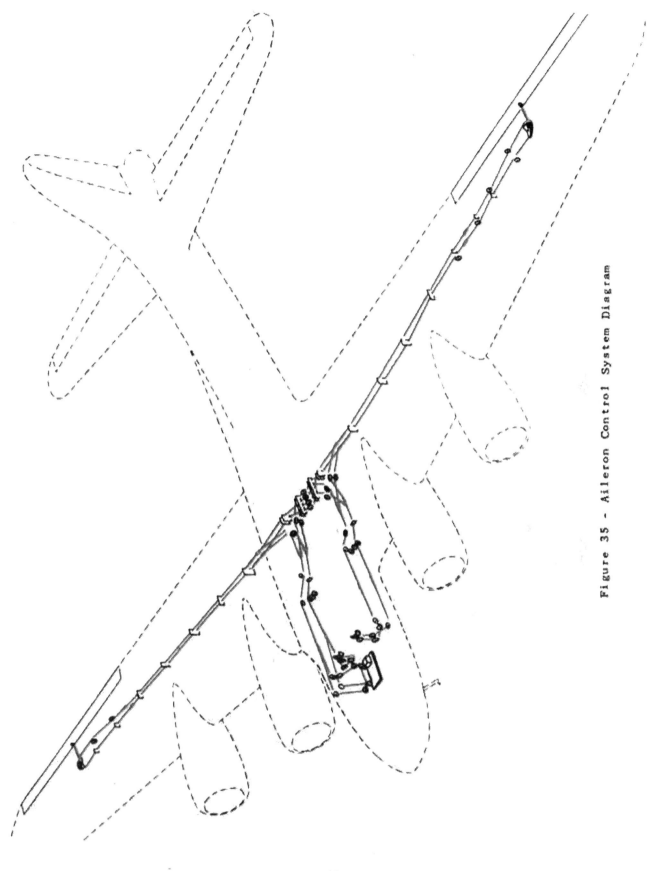

Figure 35 - Aileron Control System Diagram

SECTION II

PILOT'S COMPARTMENT INSTRUCTIONS

1. Before Entering the Pilot's Compartment.

Check and sign form F, the Weight and Balance Clearance, prepared by the ground loading personnel. This may be rapidly and accurately accomplished by using a load adjuster. Instructions and sample loading problem are published in paragraph 4. of section I as condensed instructions for the information and guidance of all personnel using a load adjuster to determine change in balance from the basic airplane to the loaded airplane as flown, and to insure that the weight distribution of all items loaded above and beyond the basic airplane weight and balance will not produce a weight and balance condition beyond permissible limits.

2. On Entering the Pilot's Compartment.

   a. Special Check for Night Flying.

     (1) Master battery switches (figure 5-16) "ON."

     (2) Turn control panel lights (figure 39-7) "ON."

     (3) Turn side control panel lights (figure 5-5) "ON."

     (4) Test operate the instrument panel lights.

     (5) Test operate the landing lights. (See figure 39-9.)

WARNING: Do not permit lights to burn more than five seconds during test.

     (6) Test operate the identification lights. (See figure 39-4.)

     (7) Test operate the passing lights. (See figure 5-9.)

     (8) Test operate the position lights. (See figure 5-17.)

Figure 36

Pilot's Instrument Panel

1 AC Voltmeter
2 Emergency Oil Pressure Warning Lamp
3 Hydraulic Oil Pressure Gage
4 Oil Pressure Warning Lamp
5 Bomb Release Warning Lamp
6 Suction Gage
7 Pilot Director Indicator
8 Altimeter Correction Card
9 Aileron Locking Pin
10 Ultra-Violet Spotlight Controls
11 Static Pressure Selector Valve

1 Panel Light	13 Turn Indicator
2 S.C.R. 535 Destruction Switches	14 Propeller 3 & 4 Feathering Button
3 Radio Compass Indicator	15 Bank & Turn Indicator
4 S.C.R. 535 Remote Control Switch	16 Flight Indicator
5 Windshield Defroster Control	17 Rate of Climb Indicator
6 Ultra Violet Spot Light	18 Bomber Call Lamp
7 Panel Light	19 Landing Gear Warning Lamp
8 " " Switch	20 Tail Wheel Lock Warning Lamp
9 Altimeter	21 Manifold Pressure Gage for Eng. 1 & 2
10 Propeller 1 & 2 Feathering Buttons	22 Tachometers
11 Marker Beacon Indicator Lamp	23 Manifold Pressure Gage for Engines 3 & 4
12 Air Speed Indicator	

Figure 37 - Pilot's and Copilot's Instrument Panel

b. Check for All Flights.

PILOT

(1) Emergency ignition switch (figure 39-11) "ON."

(2) Check each battery switch (figure 5-16) separately with either inverter on.

(3) Master battery switches (figure 5-16) "ON."

(4) Turn hydraulic pump switch "ON." If it is momentary "AUTO-MANUAL" type, it should remain in "AUTO" unless the pump fails to operate.

(5) Landing gear control switch (figure 39-8) in neutral.

(6) Flap control switch (figure 39-10) in neutral.

(7) Have copilot set parking brake.

(8) Ascertain free movement of flight control column (figure 40-8), wheel and rudder pedals to the extremities of their operating ranges.

COPILOT

(7) Set parking brake (figure 38-15) at command of pilot.

3. Starting the Engines.

   a. If the engines have stood for over two hours, have the propellers turned over three complete revolutions by hand. Be sure ignition switches (figure 39-1) are "OFF."

   b. Order aerial engineer to open manual shut-off valve and set selective check valve (figure 66) to "SERVICING" position.

   c. Check hydraulic pressure, both accumulators (600-800 pounds per square inch). Order top gunner to close manual shut-off valve. Set selective check valve to "NORMAL" position.

   d. Cabin heat control (figure 42-10) in "OFF" position.

   d. Open cowl flaps and return valves (figure 39-12) to "LOCKED" position.

   e. Move turbo controls (figure 40-2) to "OFF."

   e. Fuel transfer valves (figure 70-2) and pump switch (figure 70-1) should be "OFF." Have aerial engineer check them.

   f. Open all fuel shut-off valves. (See figure 39-3.)

   f. Set fire extinguisher selector valve (figure 38-16) to engine being started. Auxiliary external extinguishers should be available nearby.

   g. Crack throttles (approximately 1000 rpm).

   g. Move intercooler controls (figure 44-9) to "COLD."

   h. Direct copilot to open carburetor air filters.

   h. Open carburetor air filters (figure 38-10) when directed by pilot.

   i. Set propeller controls (figure 40-3) for high rpm.

   i. Move mixture controls (figure 40-7) to "ENGINE OFF."

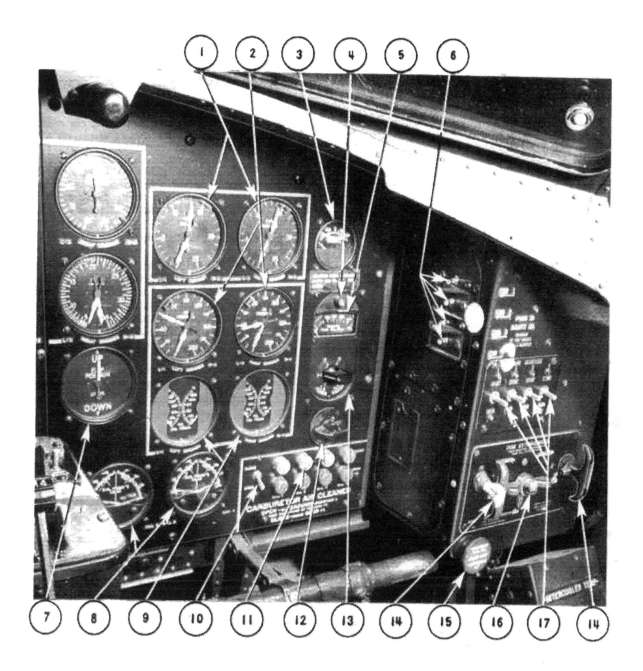

1. Fuel Pressure Gages
2. Oil Pressure Gages
3. Free Air Thermometer
4. Fuel Supply Warning Lamp
5. Liquidometer Indicator
6. Oil Dilution Switches
7. Flap Position Indicator
8. Cylinder Head Temp. Gages
9. Oil Temperature Gages
10. Carburetor Air Cleaner Switch
11. Carburetor Air Cleaner Warning Lamps
12. Ultra-Violet Spot Light Control
13. Fuel Tank Selector Valve
14. Fire Extinguisher Discharge Handle
15. Parking Brake Locking Handle
16. Engine Fire Extinguisher Selector Valve
17. Engine Starting Switches

Figure 38 - Copilot's Instrument & Auxiliary Panels

PILOT

j. Turn switch for engine affected (figure 39-1) to "BOTH."

m. Direct copilot to start engines. Recommended starting order is 1-2-3-4.

p. When the engine fires, move the mixture control (figure 40-7) to "AUTOMATIC RICH."

CAUTION: Do not advance the throttles as lean mixture and backfire hazard will result.

t. If no oil pressure is indicated within one-half minute after starting, direct copilot to stop engine with mixture control. Cut ignition and investigate.

u. In case of fire in a nacelle, run up the engine in an attempt to blow out the fire. If this fails, direct copilot to stop the engine.

v. Close cowl flaps if the fire is in nacelle 1 or 2.

w. If fire is not smothered by closing the cowl flaps, close fuel shut-off valve (figure 39-3), stop booster pump, and direct copilot to pull fire extinguisher, both charges if necessary.

NOTE: If engine accessory cowling is not installed, it is unlikely that the fire can be extinguished by the CO_2 system. External fire extinguishers must therefore be used.

COPILOT

j. Set primer (figure 44-10) to "OFF" position.

k. Start No. 3 fuel booster pump for primer pressure. It should be six-eight pounds per square inch.

l. Start fuel booster pump for engine affected.

m. When directed by pilot, move starter switch (figure 38-17) of engine affected (recommended starting order is 1-2-3-4) to "START" position and hold for approximately 30 seconds.

n. While starter switch is in "START" position, unlock primer (figure 44-10), set to engine affected, and expel air from line by pumping until a solid charge of fuel is obtained.

o. When directed by pilot, move starter switch to "MESH" position.

p. When the starter is meshed, prime with quick strokes (to atomize the primer charge) until the engine fires.

q. If necessary to prevent engine from quitting due to lack of fuel, pump primer with several slow strokes.

CAUTION: Return primer (figure 44-10) to "OFF" position.

r. Shut off booster pump if fuel pressure from engine pump remains steady.

s. If engine stops, return mixture control to "ENGINE OFF" immediately, cut ignition switch and repeat the starting procedure.

t. After engine starts, check for indication of oil pressure. Notify pilot if no pressure is indicated within one-half minute. Move mixture control to "ENGINE OFF" when directed by pilot.

u. When directed by pilot, stop engine by moving mixture control (figure 40-7) to "ENGINE OFF."

v. Close cowl flaps if the fire is in nacelle 3 or 4.

w. Pull fire extinguisher charges at command from pilot.

RESTRICTED   T.O. NO. 01-20EF-1

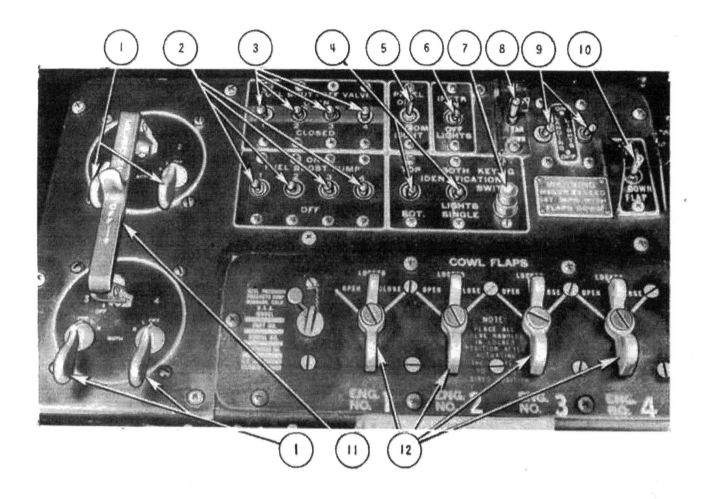

1	Ignition Switches	7	Identification Lights Keying Switch
2	Fuel Boost Pump Switches	8	Landing Gear Control Switch
3	Fuel Shut-off Valve Switches	9	Landing Lights Switches
4	Identification Lights Switches	10	Wing Flaps Control Switch
5	Panel Light Switch	11	Emergency Ignition Switch
6	Fluorescent Lamps Switch	12	Cowl Flap Control Valves

Figure 39 - Pilot's Central Control Panel

RESTRICTED   T. O. No. 01-20EF-1

PILOT   COPILOT

x. Before resuming operations after fire, be sure that CO₂ cylinders are replaced.

4. **Engine Warm-up.**

a. When oil temperature begins to rise and oil pressure is 50 pounds per square inch, open throttles to 1000-1250 rpm.	a. Notify pilot when oil temperature begins to rise and oil pressure is 50 pounds per square inch.
b. When engines are thoroughly warmed, the rpm may be increased for instrument check.	b. Notify pilot when maximum temperature and pressure values are reached.

CAUTION: 2500 rpm must not be maintained for more than one-half minute and the following values must not be exceeded:

Fuel Pressure   16 lb/sq in.
Oil Pressure   80 lb/sq in.
Oil Temperature   88°C (190.4°F)
Cylinder Temp.   205°C (401°F)

5. **Emergency Take-off.**

a. If the airplane has been on the "alert," the engines will have been started, and will be warm and ready for take-off by the time the flight crew gets within the airplane. The pilot will proceed with a routine take-off, being careful not to exceed 46 inches Hg manifold pressure.

b. If an emergency take-off is necessary with cold engines, due to the lack of a ground crew, the following procedure should be followed:

(1) Start engines, using oil dilution as soon as engines fire in order to get minimum oil pressure of 70 pounds per square inch.

(2) Fuel pressure should be at least 12 pounds per square inch.

(3) Set wing flaps for take-off, leave cowl flaps less than one-third open to expedite warm-up. Proceed with take-off. Do not exceed 46 inches Hg manifold pressure.

6. **Engine and Accessories Ground Test.**

PILOT   COPILOT

a. Direct gunner to secure lower turret with guns pointing rearward.	a. See that all doors and hatches are closed.
b. Set altimeter.	b. Hydraulic pressure should be 600 to 800 pounds per square inch on each gage.
c. Turn on master ignition switch (figure 39-11), battery switches (figure 5-16), and the hydraulic switch (figure 5-21). ("AUTO-MANUAL" type in "AUTO" position.)	c. Warning and indicator lights should be: Tail wheel unlocked - On (red) Landing gear - On (green) Hydraulic pressure: Service - Off. Emergency - Off. Vacuum - Off.
d. A.F.C.E. switches, (figure 41-1) "OFF," all knobs on control panel, "POINTERS-UP," turn control, "CENTERED."	d. Check all fuel quantities.
e. Set propeller controls for high rpm and lock.	e. Set intercooler controls (figure 44-9) to "COLD" unless icing conditions exist.

Revised 4-1-43   RESTRICTED

1 Throttle Control Lock
2 Turbo Supercharger Controls
3 Propeller Controls
4 Propeller Control Lock
5 Turbo & Mixture Controls Lock
6 Throttles
7 Mixture Controls
8 Flight Control Column

Figure 40 - Power Plant Controls

PILOT

<u>f</u>. Flight controls unlocked. Move them to the limits of their ranges to insure free operation.

<u>i</u>. Contact control tower for clearance.

<u>j</u>. Signal ground crew to remove wheel chocks.

<u>k</u>. With mixture controls (figure 40-7) in the "AUTOMATIC RICH," check ignition at 1500 to 1600 rpm.

<u>l</u>. Check propeller governor at 1500 rpm by moving control to high pitch. When rpm decreases to approximately 1100, return control to low pitch position and lock.

<u>m</u>. Run up each engine individually and adjust supercharger regulator control stops for 46 inches Hg manifold pressure at full throttle and 2500 rpm.

IMPORTANT: This adjustment must be made as quickly as possible and must not exceed one-half minute for each engine.

<u>n</u>. Set trim tabs in neutral.

<u>o</u>. Check flight controls.

WARNING: Operate to full extent of their ranges to insure free and proper movement.

<u>p</u>. Close window.

. Taxiing.

<u>a</u>. Inboard throttles may be locked for taxiing with outboard engines.

COPILOT

<u>f</u>. Cowl flaps should be open. Check visually.

<u>g</u>. Wing flaps up. Switch (figure 39-10) in neutral.

<u>h</u>. Tail wheel unlocked. Locking handle (figure 41-5) should be in up position.

<u>k</u>. Check the following during ignition check:

Fuel Pressure: Desired - 12-16 lb/sq in.
Maximum - 16 lb/sq in.
Minimum - 12 lb/sq in.

Oil Pressure: Desired - 75 lb/sq in.
80 lb/sq in.
70 lb/sq in.

Oil Temp: Desired - 70°C (158°F)
Maximum - 88°C (190°F)
Minimum - 60°C (140°F)

Cylinder Temp: - 205°C (401°F) Maximum

<u>m</u>. Notify pilot if any temperature or pressure reading is not satisfactory.

<u>o</u>. Turn all fuel boost pumps "ON."

<u>p</u>. Close window.

<u>a</u>. Notify pilot if:

Cylinder temperature exceeds 205°C (401°F).
Oil pressure exceeds 75 pounds per square inch or is less than 15 pounds per square inch for idling engines.
Oil inlet temperatures exceeds 70°C (158°F).
Fuel pressure is over 16 pounds per square inch or under 12 pounds per square inch.

<u>b</u>. Lock tail wheel (warning lamps off) after airplane has taxied to take-off position.

1. Automatic Flight Control Equipment Switches
2. Rudder and Elevator Lock Control Handle
3. Rudder Trim Tab Control Wheel
4. Rudder Trim Tab Indicator
5. Tail Wheel Lock

Figure 41 - Pilot's And Copilot's Lower Control Stand

PILOT   COPILOT

8. Take-off.

   a. Refer to the FLIGHT OPERATION INSTRUCTION CHARTS, section III for all flight operation and engine operating data.

   b. Turn generator switches (figure 5-8) "ON."

   c. Open throttles slowly to FULL THROTTLE (three to five seconds).

   d. With a runaway turbo or propeller, follow the following instructions:

   (1) THROTTLE BACK FIRST.

   (2) Move turbo control (figure 40-2) to "OFF."

   (3) If necessary, set propeller controls (figure 40-3) in "LOW RPM." There is small likelihood of a runaway turbo, but the danger is great if it occurs during a take-off. The pilot MUST be alert during the take-off to note immediately and correct any excessive manifold pressure.

   e. When airplane is clear of the ground, direct copilot to retract the landing gear.

   e. Retract landing gear at command from pilot.

   f. Accelerate to speed for cruising climb.

   f. Cylinder head temperatures must not exceed 260°C (500°F) (five minutes maximum).

   Oil pressure - desired - 80 lb/sq in.
   Oil Temp - desired - 70°C (158°F)
   Fuel Pressure - 12-16 lb/sq in.

   g. Adjust intercooler control (figure 44-9) to "COLD" unless icing conditions prevail.

9. Engine Failure During Take-off.

   a. Failure of an engine during take-off may not be noticeable immediately except for a resultant swing. If, therefore, a swing develops, and there is room to close the throttles and pull up, this should be done.

   b. If it is necessary to continue with the take-off, even though one engine has failed, hold the airplane straight by immediate application of rudder. Gain speed as rapidly as possible. See that the landing gear is up, or coming up, and feather the propeller of the dead engine. For detailed instructions on feathering and unfeathering the propellers, refer to paragraphs 13. and 14. of this section. Retrim as necessary.

Revised 4-1-43   RESTRICTED

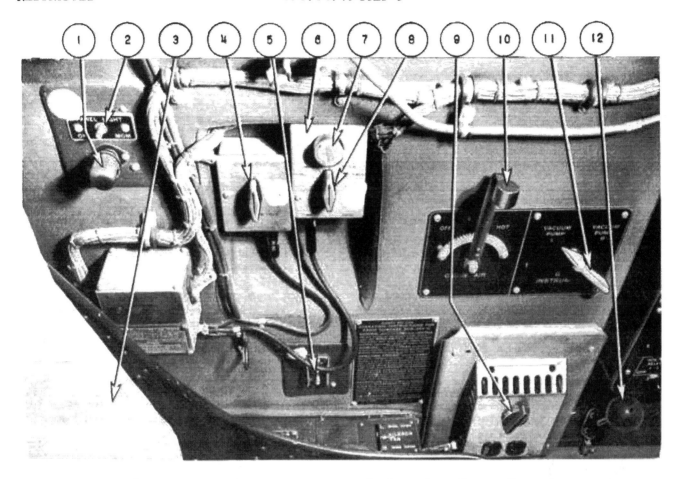

1. Panel Light
2. Panel Light Switch
3. Pilot's Seat
4. Crystal Filter Selector Switch
5. Propeller Anti-icer Switch
6. Interphone Jackbox
7. Interphone Volume Control
8. Interphone Selector Control
9. Suit Heater Outlet
10. Cabin Air Control
11. Vacuum Pump Selector Control
12. Emergency Bomb Release

Figure 42 - Pilot's Controls - Left Side Wall

Figure 43 - Floor Panel Left of Pilot's Seat

1. Propeller Anti-Icer Switch
2. Propeller Anti-Icer Rheostat Controls
3. Wing De-icer Control Valve
4. Aileron Trim Tab Position Indicator
5. Aileron Trim Tab Control
6. Pilot's Seat Adjustment Lever

PILOT | COPILOT

10. Climb.

a. Reduce manifold pressure with supercharger controls. (See figure 40-2.)

b. Reduce rpm as required for climb. | b. Adjust cowl flaps as required to maintain proper cylinder head temperature.

c. Make a visual check of engines 1 and 2. | c. Make a visual check of engines 3 and 4.

d. Adjust trim tabs as required.

11. Flight Operation.

a. Instructions for use of the Flight Operation Instruction Charts during flight.

(1) Determine gross weight of airplane with the aid of the Weight and Balance Chart in section III. | (1) Follow the "INSTRUCTIONS" for use of the chart printed near the top of the chart.

(2) Select the Flight Operation Instruction Chart for the gross weight (including any external load item) of the airplane. | (2) A series of charts covering pertinent changes in gross or aerodynamical load (due to fuel consumption or other disposable load being dropped) will be found in their most logical sequence in section III.

b. Adjust engine controls as required. See Flight Operation Instruction Charts in section III. | b. Set mixture controls (figure 40-7 to "AUTOMATIC LEAN."

c. Order copilot to set carburetor air filter switch to "CLOSED" at 8000 feet unless dust conditions are found above that altitude. | c. When ordered by pilot, move switch (figure 38-10) to "CLOSED."

WARNING: Switch must never be left in the "OPEN" position above 15,000 feet.

d. Flight Controls

CATUION: Instantaneous load factors above the allowable can be reached very easily with rough elevator control movements. In turbulent air or in combat maneuvering, corrections should be made very smoothly.

e. Automatic Flight Controls. | e. Adjust cowl flaps as required to maintain proper cylinder head temperatures.

(1) Throw "ON" master and stabilizer switches.

(2) CAREFULLY TRIM AIRPLANE FOR STRAIGHT AND LEVEL FLIGHT.

(3) Turn "ON" tell-tale lights.

(4) After master and stabilizer switches have been "ON" for ten minutes, throw "ON" PDI and servo switches.

(5) Center PDI by turning plane and resuming straight and level flight.

(6) With PDI on "ZERO," adjust rudder centering knob until both rudder tell-tale lights go "OUT." Immediately throw rudder switch "ON."

(7) With wings level, adjust aileron centering knob until both aileron tell-tale lights go "OUT." Immediately throw aileron switch "ON."

1 Microphone "Push-to-talk" button
2 Suit Heater Outlet
3 Oxygen Regulator
4 Panel Light
5 Panel Light Switch
6 Control Column
7 Control Wheel
8 Rudder Pedal Adjustment Lever
9 Intercooler Controls
10 Primer

Figure 44 - Copilot's Controls - Right Front Side Wall

PILOT

(8) With airplane flying level, adjust elevator centering knob until both elevator tell-tale lights go "OUT." Immediately throw elevator switch "ON."

(9) Observe PDI, artificial horizon, and rate-of-climb or altimeter instruments. Then carefully retrim all centering knobs, until ship is flying as straight and level as possible, with PDI on "CENTER."

(10) With auto-pilot engaged, all course corrections must be made with turn control, ONLY. Always turn knob with a slow steady movement.

h. Cruising Climb.

(1) Reduce manifold pressure with turbo controls. (See figure 40-2.)

12. Fire in Flight.

a. Notify copilot if fire is in engine 1 or 2.

b. Attempt to blow out the fire in the nacelle by running up the engine. If this fails, direct the copilot to close the cowl flaps of the affected engine.

c. Close proper fuel shut-off valve. (See figure 39-3.)

d. Continue with throttles unchanged. Order copilot to set fire extinguisher selector valve and to stand by to pull charge.

e. If fire is not extinguished, order copilot to pull first charge.

f. If fire persists, order copilot to pull second charge.

g. Feather propeller after fire is extinguished if serious damage has resulted or if excessive vibration occurs during fire.

13. Propeller Feathering.

a. Notify copilot to stop engine affected.

b. Turn automatic flight control equipment switches (figure 41-1) "OFF."

c. Notify copilot to press proper feathering switch.

COPILOT

f. Stop booster pumps until needed (which will be above 15,000 feet).

g. Begin flight performance log and made entries in Form I as required.

h. Cruising Climb.

(1) Set mixture controls (figure 40-7) in "AUTOMATIC RICH."

(2) Adjust cowl flaps as required to maintain proper cylinder head temperatures.

a. Notify pilot if fire is in engine 3 or 4.

b. Close cowl flaps when directed by pilot.

d. Set selector valve (figure 38-16) when ordered by pilot and stand by to pull charge.

e. When ordered by pilot, pull first charge.

f. Pull second charge when directed by pilot.

CAUTION: If fire occurs in more than one engine, do not attempt to distribute a single charge because the volume of a single cylinder is insufficient for two engines.

a. Move mixture control of affected engine to "ENGINE OFF."

b. Stop the booster pump if running.

c. Press proper feathering switch. (See figure 37-10 or figure 37-14.)

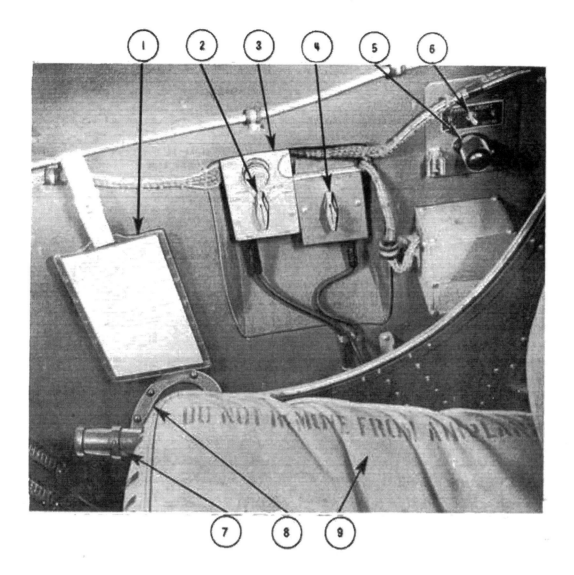

1. Check List
2. Interphone Selector Switch
3. Interphone Jackbox
4. Crystal Filter Selector Switch
5. Panel Light
6. Panel Light Switch
7. Hydraulic Hand Pump
8. Primer
9. Copilot's Seat

Figure 45 - Copilot's Controls - Right Side Wall

PILOT

d. When propeller stops, turn proper ignition switch (figure 39-1) to "ENGINE OFF."

e. Close throttle.

f. Adjust trim tabs as required.

g. Turn automatic flight control equipment switches (figure 41-1) "ON."

h. If the engine is not to be restarted, order engine fuel transferred to other tanks as required.

i. When No. 2 engine is affected:

(1) The glycol pump is inoperative. If cold air is not desired in the cabins, shut off heating and ventilating system by moving control handle (figure 42-10) fully aft.

(2) When one vacuum pump is inoperative, (engine No. 2 or No. 3): Set vacuum pump selector ("GYRO INSTR.") valve (figure 42-11) to the other vacuum pump. (De-icer pressure will thus be reduced and de-icer vacuum will not be available. De-icer system will therefore operate inefficiently.)

14. Propeller Unfeathering.

a. Notify copilot which engine is to be restarted.

b. Turn automatic flight control equipment switches (figure 41-1) "OFF."

d. Crack proper throttle to 1000 rpm approximately.

e. Turn ignition switch (figure 39-1) to "BOTH."

f. Press proper feathering switch (figures 37-10 and -14) and hold it closed until engine speed reaches 1000 rpm.

g. Open throttle slowly to 1200 rpm.

h. Adjust trim tabs as desired.

i. Maintain 1200 rpm until notified by copilot that oil temperature is 70°C (158°F).

k. Synchronize manifold pressure and rpm with other engines.

CAUTION: Above 15,000 feet, power must be adjusted with turbo control - full throttles.

l. Adjust trim tabs as required.

m. Turn automatic flight control equipment switches (figure 41-1) "ON."

NOTE: When No. 2 propeller is unfeathered, the pilot may turn on the heating and ventilating system by moving the control (figure 42-10) to any position between one-half and fully forward.

COPILOT

d. Close cowl flaps of engine affected.

h. Assist aerial engineer to transfer fuel from the dead engine tank.

a. Set propeller control (figure 40-3) to "LOW" rpm.

b. Set intercooler control (figure 44-9) to "HOT" position.

c. Close cowl flaps.

d. Start proper booster pump (if above 15,000 feet).

e. Check fuel quantity in proper tank.

f. When engine speed reaches 1000 rpm, move mixture control from "ENGINE OFF" to "AUTOMATIC RICH."

i. Notify pilot when oil temperature reaches 70°C (158°F).

j. When cylinder head temperature reaches 205°C (401°F), open cowl flaps as required for continuous operation.

k. Adjust intercooler control (figure 44-9) as required.

Revised 4-1-43

15. **General Flying Characteristics.**

   a. **General Stability.**

   (1) Increasing the power on the inboard engines causes the airplane to become slightly tail heavy, while a change of power on the outboard engines has no appreciable effect upon the trim.

   (2) Closing the cowl flaps on the inboard engines causes a similar tail heaviness, but cowl flaps on the outboard engines have a negligible effect upon the trim.

   (3) With the airplane properly trimmed for a landing with power off and flaps down, the pilot may apply power, throw the flap switch into the up position and go around with no change in trim tab setting if a second approach is necessary. The flaps retract at a satisfactorily slow rate.

   b. **Take-off.** - During the take-off run, directional control should be maintained with rudder movement and throttles, differential throttling being done with the outboard engines as much as possible.

   c. **Climb.** - The airplane will require very little elevator trim and the elevator control pressure will build up rapidly as the climbing speed is reduced below normal.

   d. **Level Flight.** - In normal flight, turns can be made very smoothly with aileron control only. In instrument flight, the pilot should pay special attention to holding the wing level, because the directional stability produces a noticeable turning tendency with one wing down.

   WARNING: Care should be taken to avoid excessive use of the ailerons.

   e. **Rough Air Operation.**

   (1) The ailerons and rudder can be used without concern regarding excessive loads. It is almost impossible to damage the system without a deliberate attempt to do so. The forces required are small enough and the resultant responses large enough to maintain ample control of the airplane.

   (2) In the case of the elevators, however, care must be exercised to assure smooth operation. In thunder storms, squalls, and in or near extremely turbulent cumulous clouds, it is possible to develop excessive load factors with the elevators unless proper care is exercised.

   (3) Operation in rough air should be made on the basis of holding constant the airspeed with the elevator. Corrections for changes in altitude must be done with power, and for very rapidly rising air currents, it may be necessary to lower the landing gear.

   (4) The airplane should not be dived through a cloud layer or through rough air at the limiting diving speed of 305 mph, nor should high speed flight be attempted in rough air.

   f. **Carburetor Air Filters.**

   (1) Turning the carburetor air filter switch to "OPEN" before take-off, without adjusting the supercharger controls, decreases the manifold pressure approximately 1-1/2 inches. This drop may be overcome by readjusting the supercharger control lever stops until the desired manifold pressure is obtained.

   (2) It is recommended that the air filters remain in the "OPEN" position for all ground operations and for dust conditions up to 8000 feet. Occasionally dust is found in the atmosphere above 8000 feet, in which case it is permissible to leave the filters OPEN up to 15,000 feet, if necessary.

   (3) Use of the filters above 8000 feet is to be avoided, if possible, since any operation above that altitude is accompanied by a rise in carburetor air inlet temperature which increases the possibility of detonation. This condition would be further aggravated by abnormally high outside air temperatures. The turbo also has a tendency to overspeed, thus shortening the life of the unit. IN ALL CASES, THE FILTERS MUST BE CLOSED ABOVE 15,000 FEET. This is the critical altitude at which carburetor air inlet temperatures become dangerously high and the pressure differential across the turbine wheel becomes great enough to induce appreciable overspeeding of the turbo when the induction air is drawn through the filters. Failure to observe this precaution may cause detonation and eventual engine failure or sufficient overspeeding of the turbine wheel to seriously damage the supercharger.

   (4) It is important that the filter switch be in the "OPEN" position before landing, since the supercharger control levers were adjusted for a maximum manifold pressure of 46 inches Hg at take-off with the filters OPEN. If emergency power is attempted with the filters closed, manifold pressures above the recommended maximum of 46 inches will be obtained.

   g. **Obtaining Maximum Performance.**

   (1) The ceiling and climb at 35,000 feet are as great or greater than that of many fighter airplanes, but the high speed is not as great as most fighters at normal altitudes; therefore, in order to outperform any enemy at 35,000 feet it will be necessary to outclimb him rather than to outdistance him.

   (2) The increase of speed obtained by nosing the airplane down below the horizontal at rated power and at any high power condition is smaller than that obtained by fighters.

   (3) In order to obtain maximum climb, the following technique should be used:

   (a) Maintain the proper climbing airspeed (135 mph indicated).

Altitude	Manifold Pressures giving rated power at 2300 engine rpm and 21,300 turbo rpm		Manifold Pressures giving military power at 2500 engine rpm and 21,300 turbo rpm	
0	39.0		47 in.	
10,000	38.0		46 in.	Military Power to 28,000 ft
20,000	37.5	Rated Power	45 in.	
30,000	37.0		41.5 in.	These manifold pressures not allowable below 2500 rpm
31,000	37.0		40.0 in.	
32,000	36.5	These manifold pressures not allowable below 2300 rpm	38.5 in.	
33,000	35.0	Decreasing Power	37.0 in.	
34,000	33.5		35.0 in.	
35,000	32.0		33.0 in.	

(b) In any emergency whatever, such as being pursued by the enemy, engine speed should be increased to 2500 rpm. The increase in rpm has a very appreciable effect on increasing propeller efficiency and rate of climb under conditions of climbing speed and high altitude, and, in addition, is not detrimental to the engine. The pilot should avoid the use of less than 2500 rpm when primarily interested in a high rate of climb at high altitudes.

(c) 21,300 rpm has been determined to be the maximum operating turbo speed with a five percent overspeed allowance in emergencies. This would provide an emergency rating of 22,400 rpm. At any altitude greater than 30,000 feet and at any power obtained in automatic rich (with 2300 rpm or 2500 rpm, full throttle and turbos set for manifold pressures indicated in the following table), the exhaust gas temperatures are dropping rapidly and it is very unlikely that critical temperatures will be approached. The following tentatively determined manifold pressures will permit safe operation of the turbo under the given conditions:

NOTE: This table is based on the best present available information for maximum performance at 55,000-pound gross weight with carburetor air filters closed.

NOTE: Obviously, all four turbo installations are not identical and hence, operation according to the above table will not result in identical turbo rpm for all engines.

(d) The outboard engines have higher critical altitudes than the inboards by approximately 2000 feet to 3000 feet, and the inboard engine without boilers in the stack has a 1500-foot higher critical altitude than the engine with the boilers in the stack. The critical altitude of the outboard engines as far as limiting turbo rpm is concerned is 31,000 feet.

(e) The above table actually applies only to the outboard engines. However, the differences between the inboard and outboard engines are covered by the margin of safety incorporated in the design of the turbo itself. Even though 22,400 rpm are allowable for military power operation, the right-hand column of the above table, is made for only 21,300 rpm.

h. Landing. - During the approach for landing very little change in elevator trim will be required. As the flaps are lowered the airplane becomes slightly tail heavy, but if it is trimmed slightly nose heavy at 147 mph with flaps up, it will be properly trimmed at 120 mph with flaps down. This is a satisfactory approach speed for gross weights below 50,000 pounds.

16. Stalls.

a. Stalling characteristics are very satisfactory. Under no condition is there any sharp tendency to roll. Yawing is sufficiently suppressed to make any rolling at the stall of a very mild nature. Under all conditions a stall warning of several miles per hour is indicated by buffeting of the elevators.

b. A pitching motion started by the elevators should be damped slowly. It will easily reduce the airspeed well below the stall unless it is deliberately stopped.

c. Full flap reduces the stalling speed about 15 mph for gross weights between 40,000 and 45,000 pounds, but full military power for the same loading conditions may reduce the stalling speed another 15 mph. Accidental or deliberate yawing will increase the stalling speed and increase any tendency to roll at the stall.

d. The ailerons have a tendency to overbalance and reverse effectiveness at the stall. For example, if the left wing tends to drop at the stall and right aileron control is applied in an attempt to raise the left wing, the aileron operating forces will tend to decrease and cause full aileron deflection, but the response will be an increase in the roll to the left. THE PROCEDURE IN RECOVERING FROM A STALL IS TO HOLD THE AILERONS NEUTRAL AND REFRAIN ENTIRELY FROM THEIR USE.

e. Procedure for recovering from a stall is normal. The airspeed for normal flight must first be regained by smooth operation of the elevators. This may put the airplane into a dive of 30 degrees or less. During the process of regaining airspeed the rudder may be used to maintain laterally level flight for lateral control, but not until the airspeed is regained. RECOVERY FROM THE DIVE MUST BE DONE IN A SMOOTH MANNER. Failure to make a smooth recovery may be a restalling of the airplane or a structural failure, both due to excessive load factors.

f. Airspeed increase necessary to regain normal flight need not generally be more than 20 mph, and possibly, after practice, even less.

17. Spins.

Inadvertent spinning is very unlikely, as stability and damping are very high. The airplane is not designed for spinning, and this maneuver should never be attempted.

18. Dives.

a. The structural factors limiting the diving speed to 305 mph are the engine ring cowl strength, the wing leading edge de-icer boot strength, the pilot's compartment windshield and enclosure strength and the critical flutter speed. The engine ring cowl has been designed to withstand 420 mph. Windshield and pilot's compartment enclosure have ample margin at 305 mph. The wing leading edge de-icer boots begin to rise slightly from the wing at 305 mph, and any excessive additional speed would probably lift the upper portion of the boot well above the wing surface and allow it to flap severely against the leading edge, thus causing a structural failure.

b. When diving, it is essential that the sensitivity of the elevator trim tab be kept constantly in mind. In making dives the elevator trim tabs must be set during the dive to maintain zero elevator force and must be used with great care during recovery.

INDICATED AIR SPEED... WING FLAP SWITCH... WING FLAPS... LANDING!

19. Approach and Landing.

PILOT

a. Check cg location for landing by means of the load adjuster.

b. Set altimeter to airport pressure altitude.

c. Notify radio operator to retract trailing antenna.

d. Turn automatic flight control equipment switches (figure 41-1) off.

e. Direct copilot to adjust carburetor air filters.

COPILOT

a. SELECTIVE CHECK VALVE (figure 66) MUST BE IN "NORMAL" position.

b. Set mixture controls (figure 40-7) in "AUTOMATIC RICH."

c. Set intercooler controls (figure 44-9) in "COLD," unless icing conditions exist.

d. Radio control tower or landing clearance. Complete operating instructions for radio equipment will be found in section I, paragraph 4.i.

e. When directed by pilot, throw carburetor air filter switch (figure 38-10) to "OPEN."

PILOT

COPILOT

f. Move supercharger controls (figure 40-2) to full "ON," and propeller controls (figure 40-3) to "MAX. CRUISE." (2100 RPM.)

g. Shut off de-icer system, if operating.	g. Check instruments.
h. Order copilot to extend landing gear.	h. Extend landing gear when directed by pilot (green signal light on).
i. Check position of ball turret. Guns should be horizontal and pointing rearward.	i. Tail wheel should be locked (warning light off), locking lever (figure 41-5) flush with floor.
j. Check hydraulic pressure; it should be 600 to 800 pounds per square inch on both gages.	
k. Operate brakes. Hydraulic pressure should remain above 600 pounds per square inch. If main brakes are inoperative, prepare for emergency landing.	
	l. Check cowl flap valves. (See figure 39-12.) They must be in "LOCKED" position to guard against loss of oil supply through leaks in cowl flap actuating mechanisms.
m. After speed has dropped below 147 mph, order copilot to lower wing flaps.	m. Lower wing flaps when directed by pilot.
n. Adjust trim tabs as required.	
o. Order copilot to call off airspeed as required.	o. Call off airspeeds when directed by pilot.

20. Emergency Take-off if Landing is not Completed.

   a. Open throttle wide.

   CAUTION: Do not exceed 46 inch Hg manifold pressure.

   Raise your wing flaps before you taxi in!

   b. Increase propeller speed to 2500 rpm if controls have not been set for maximum rpm during approach as described in paragraph 15. of this section.

c. Order copilot to raise landing gear and proceed with a normal take-off.	c. Raise landing gear when directed by pilot.
d. Order copilot to raise wing flaps after 500 feet altitude has been reached.	d. Raise wing flaps when directed by pilot. (See figure 39-10.)

21. After Landing.

a. Move supercharger controls (figure 40-2) to "OFF" position.	a. Raise wing flaps. (See figure 39-10.)
b. Generator switches (figure 5-8) "OFF."	b. Check cowl flaps (open).

PILOT

c. Order tail wheel unlocked after taxi speed has dropped below 30 mph.

22. Stopping of Engines.

a. If parking brakes are set, do not permit them to remain so for very long if the brake drums are hot.

b. Idle engines at approximately 800 rpm until cylinder temperature gages (figure 38-8) show temperatures are 170°C (338°F).

c. If the airpland is to remain outside overnight, or if an engine start is anticipated in temperatures below 0°C (32°F), order copilot to dilute oil for four minutes maximum. During oil dilution period, operate supercharger controls continuously full open to fully closed in cycles of approximately 10 seconds.

d. Set propeller controls (figure 40-3) in "HIGH RPM."

e. Before stopping engines, run at 1200 rpm for 30 seconds. Direct copilot to stop engines with mixture control.

23. Before Leaving the Pilot's Compartment.

Cut off all radio, de-icer, compartment, central control panel, and pilot's side control panel switches.

24. Maneuvers Prohibited.

Spin
Loop
Roll
Inverted flight
Stall
Maximum speeds:
  Diving (normal load - flaps up)   305 mph IAS
  Flaps down   147 mph IAS
  Taxiing (tail wheel unlocked)   30 mph
Maximum manifold pressure   46 inches Hg

25. Precautions With Maximum Load.

B-17F airplanes, with modified landing gear and added chordwise wing tip tanks, can be flown up to and including a gross weight of 64,500 pounds, with the following restrictions:

At 64,500 pounds, the extra wing tip tanks must be full to obtain the effect of a relieving load on the wings in flight. Care must be exercised in taxiing; avoiding rough ground. Take-offs, above a gross weight of 56,000 pounds may be made only on smooth fields or prepared runways. All pivot turns on one wheel, while taxiing, will be avoided.

COPILOT

c. Unlock tail wheel when directed by pilot (lever, figure 41-5, as nearly vertical as possible).

c. Close oil dilution switches (figure 38-6) when ordered by pilot.

e. When directed by pilot, stop engines by moving mixture controls (figure 40-7) to "ENGINE OFF."

Complete Form 1.

All B-17 type airplanes, equipped with extra wing tip chordwise tanks, must be operated in accordance with 2., 3., and 4. above, whenever the wing tip tanks are more than half full. Maximum permissible indicated air speed of B-17F airplanes, with extra wing tip tanks full, must be limited to 230 mph, when loaded to 64,500 pounds. Maximum maneuver permissible at 64,500 pounds; positive, 2.056; negative, 1.22; landing gear, 2.1.

26. Precautions With 1600-Pound Bombs.

Some B-17F airplanes do not have a complete set of B-10 bomb shackles. 1600-pound bombs may be carried on the B-7 bomb shackle with these restrictions: If an airplane returns to base with 1600-pound bombs remaining on the racks, they shall be released, in the safe condition, over water or the safest available area. The maximum permissible gross weight of the airplane will not be exceeded when carrying 1600-pound bombs. The pilot will guard against any severe maneuvering of airplane.

SECTION III

FLIGHT OPERATION DATA

1. **Determining Gross Weight.**

Secure gross weight from Form F, Weight and Balance Clearance.

2. **Flight Planning.**

a. The following procedure may be used as a guide to assist personnel in the use of the FLIGHT OPERATION INSTRUCTION CHART for flight planning purposes.

b. If the flight plan calls for a continuous flight where the desired cruising power and air speed are reasonably constant after take-off and climb to 5000 feet, the fuel required and the flight time may be computed as a "single section" flight.

(1) Within the limits of the airplane, the fuel required and flying time for a given mission depend largely upon the speed desired. With all other factors remaining equal in an airplane, speed is obtained at a sacrifice of range, and range is obtained at a sacrifice of speed. The speed is usually determined after considering the urgency of the flight plotted against the range required. The time of take-off is adjusted so as to have the flight arrive at its destination at the predetermined time.

(2) Select the FLIGHT OPERATION INSTRUCTION CHART for the gross weight to be used at take-off. Locate the largest figure entered under G.P.H. (gallons per hour) in column I on the lower half of the chart. Add the resulting figure to the number of gallons set forth in footnote No. 2, and subtract the total from the amount of fuel in the airplane prior to starting the engines. The figure obtained as a result of the computation will represent the amount of gasoline available and applicable for flight planning purposes on the "RANGE IN AIR MILES" section of the chart.

(3) Select a figure in the fuel column equal to, or the next entry less than, the available amount of fuel in the airplane as determined in paragraph 2.b.(2) above. Move horizontally to the right or left and select a figure equal to, or the next entry greater than, the air miles (with no wind) to be flown. Operating data contained in the column number in which this figure appears represent the highest cruising speed possible at the range desired; however, the airplane may be operated in accordance with values contained under "OPERATING DATA" in any column of a higher number with the flight plan being completed at a sacrifice of speed, but at an increase in fuel economy.

(4) Using the same column number selected by application of instructions contained in paragraph 2.b.(3) above, determine the indicated air speed (in mph or knots, whichever is applicable to the calibration of the instruments in the airplane) and gallons per hour listed at sea level in the lower section of the chart under the title "OPERATING DATA." Divide this IAS into the air miles to be flown and obtain the calculated flight duration in minutes, which can be converted into hours and minutes and deducted from the desired arrival time at the destination in order to obtain the take-off time (without consideration for wind). To allow for wind, use the above IAS as ground speed and calculate a new corrected ground speed with the aid of a flight calculator or by means of a navigator's triangle of velocities.

(5) The airplane and engine operating values listed under "OPERATING DATA" in any single numbered column are calculated to give constant miles per gallon at any altitude listed. Therefore, the airplane may be operated at any altitude and at the corresponding set of values given, so long as they are in the same column that lists the range desired.

CAUTION: Ranges listed in column I under "MAX. CONT. POWER" are correct only at the altitude given in footnote No. 1, and the engine and airplane operating data listed under "OPERATING DATA" will give constant miles per gallon if operation is consistent with values set opposite the listed altitudes.

(6) The flight plan may be readily changed at any time enroute, by following the "INSTRUCTIONS FOR USING CHART" printed on each chart. The chart will show the balance of range at various cruising powers.

(7) Multiple charts are provided to give accurate data for operation at different gross weights, different external load conditions, and/or different combination of engine use, such as two or three engine operation. Extreme caution should be exercised to assure the selection of the correct chart applicable to the specific operating condition.

c. If the original flight plan calls for a mission requiring changes in power, speed, gross load, or external load, in accordance with the titles shown at the top of each chart provided, the total flight should be broken down into a series of individual short flights, each computed as outlined in paragraph 2.b. and then added together to make up the total flight and its requirement.

SUPPLEMENTARY CRUISING DATA AND CURVES

1. <u>Instructions for Use of Cruising Control Curves</u>.

   a. <u>General</u>. - Follow arrows through points A, B, C, D, E, and F to find true and indicated airspeed from temperature, pressure altitude, percent power, and gross weight. Point C is power required to make good speed at Point F with wieght of Point D. The curves are attached to the armor plate on the rear of the pilot's and copilot's seats.

   b. <u>For Use in Cruising Flight</u>. - Set manifold pressure and rpm to charted values as required to give speed or range desired. Determine density altitude, observe indicated airspeed. At charted manifold pressure and rpm in hot weather IAS will be low; in cold, high, when compared to charted values. Jockey power slightly as required (increase MP to increase speed, decrease MP to decrease speed) until charted IAS is obtained. This will establish power exactly. Fuel flow will thereby be established. Do not increase manifold pressure more than three inches above charted values without raising rpm. DO NOT EXCEED 30 inches MP or 2100 rpm for continuous cruising in auto-lean.

   > NOTE: For steady cruising it should not be necessary to vary power oftener than each hour; every three hours will probably be satisfactory.

   c. <u>Take-off Power</u>. - Take-off power is 1200 HP per engine at 2500 rpm at 46 inches manifold pressure.

   > NOTE: For long range cruising, check speed vs gross weight chart in lower left-hand corner. Speeds on this chart are for best range cruising with headwind. Refer to long range cruising charts for operating procedure for best range.

   d. <u>Mixture</u>. - Auto-lean for 750 HP and less (below 2100 rpm: 30 inches Hg). Auto-rich above 750 HP (above 2100 rpm: 30 inches MP).

   e. <u>Cowl Flaps</u>. - Closed, or as cylinder temperature permits.

   f. <u>Fuel Flow</u>. - All fuel flows are gallons per hour for <u>four</u> engines.

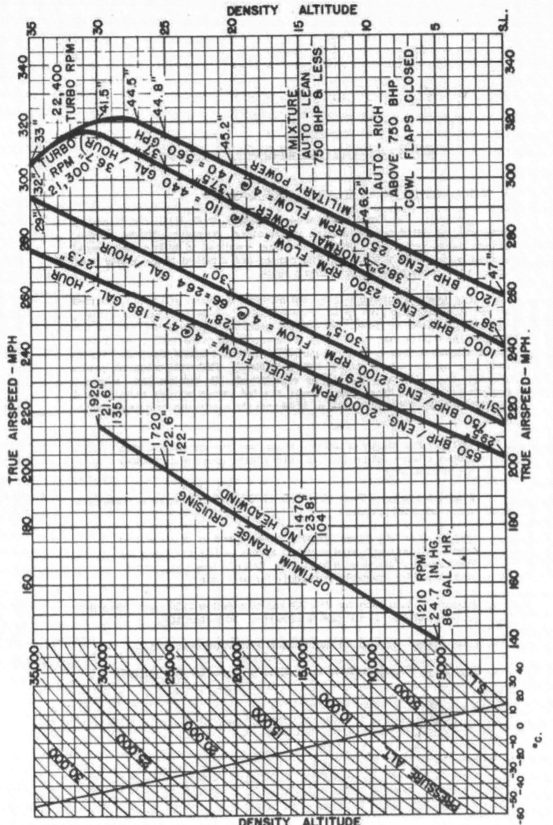

Cruising Control Chart - 35,000 Lbs.Gr.Wt.

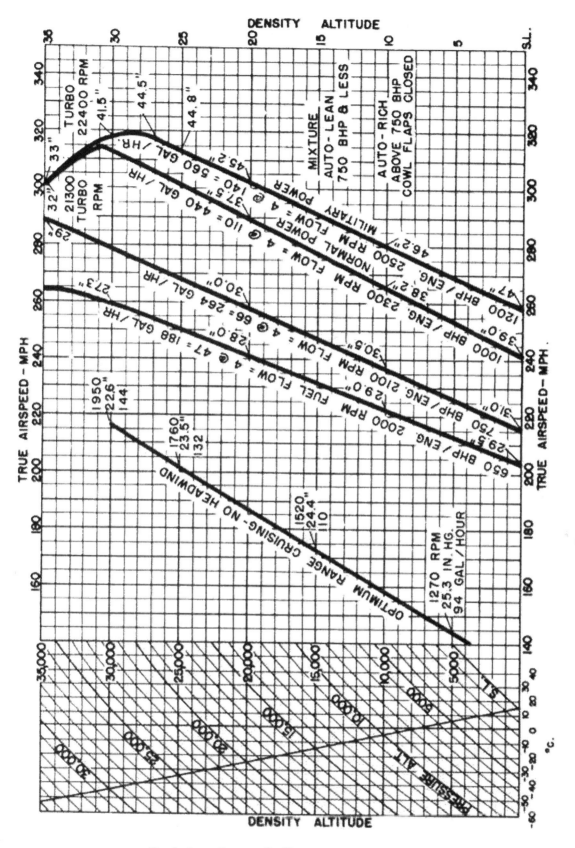

Cruising Control Chart - 37,500 Lbs.Gr.Wt.

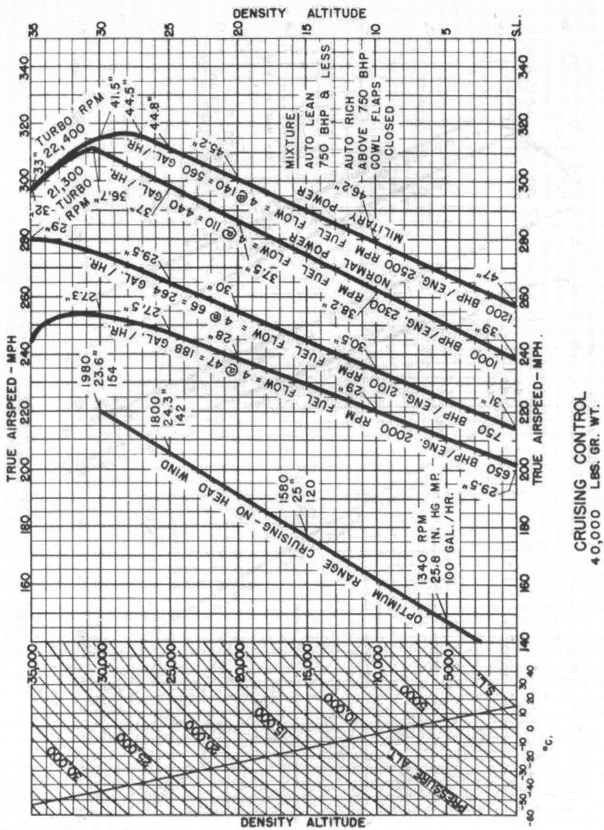

Cruising Control Chart - 40,000 Lbs. Gr. Wt.

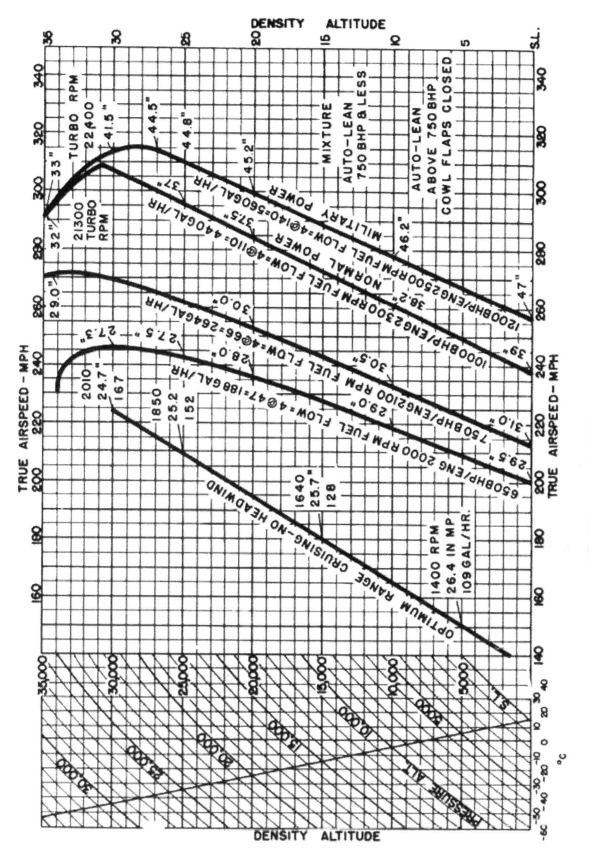

Cruising Control Chart - 42,500 Lbs.Gr.Wt.

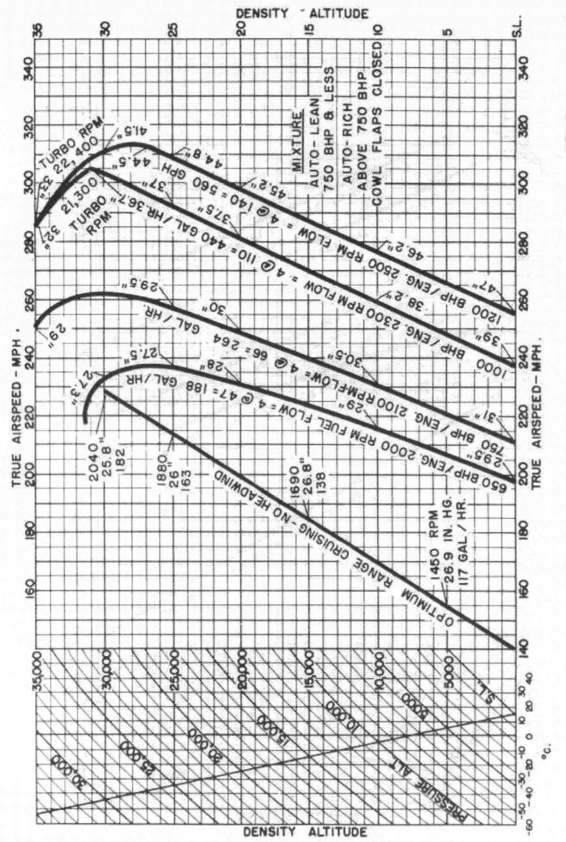

Cruising Control Chart - 45,000 Lbs. Gr. Wt.

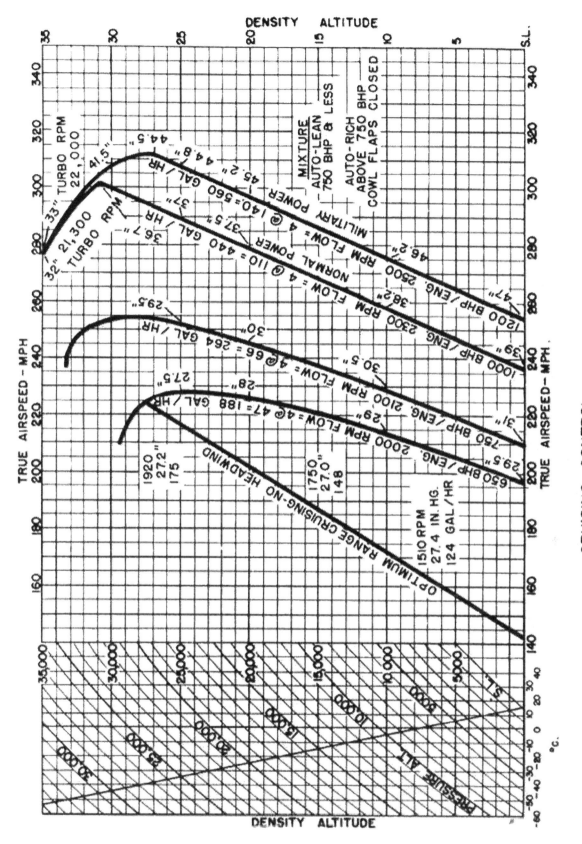

Cruising Control Chart - 47,500 Lbs.Gr.Wt.

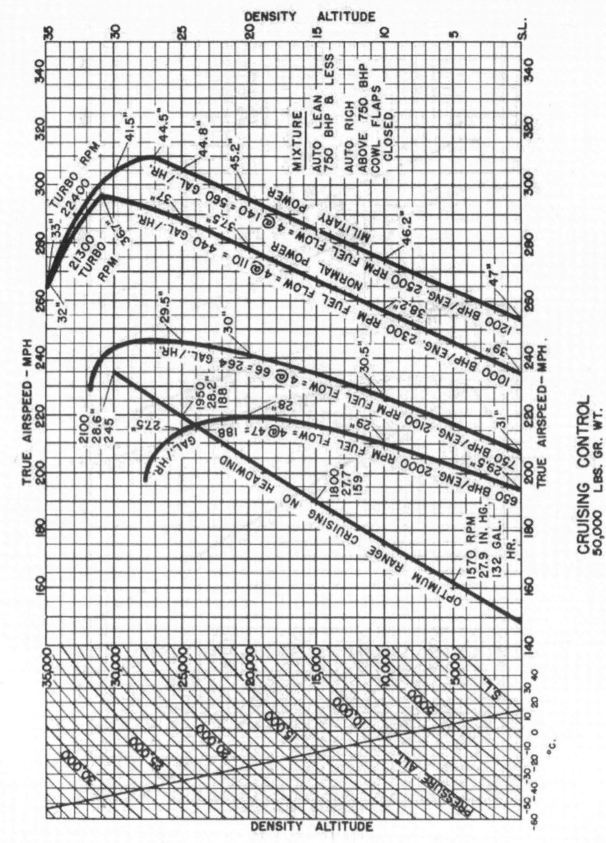

Cruising Control Chart - 50,000 Lbs.Gr.Wt.

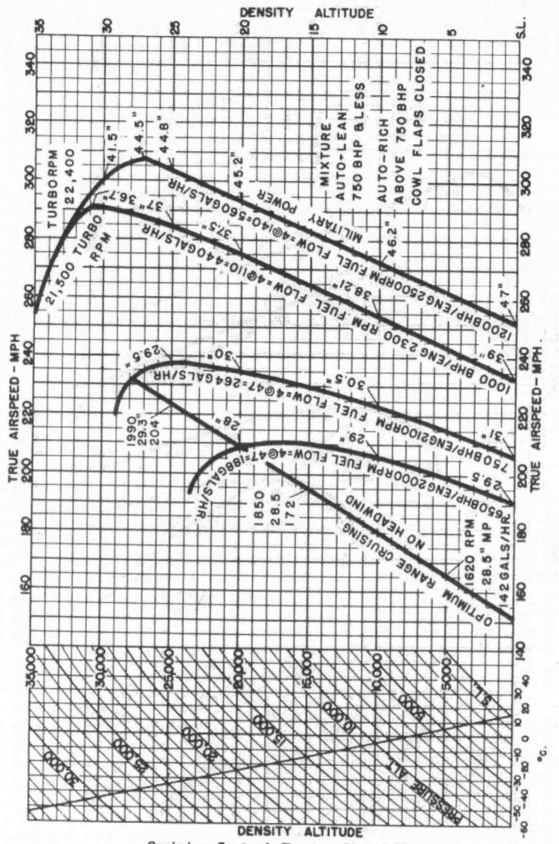

Cruising Control Chart - 52,500 Lbs. Gr. Wt.

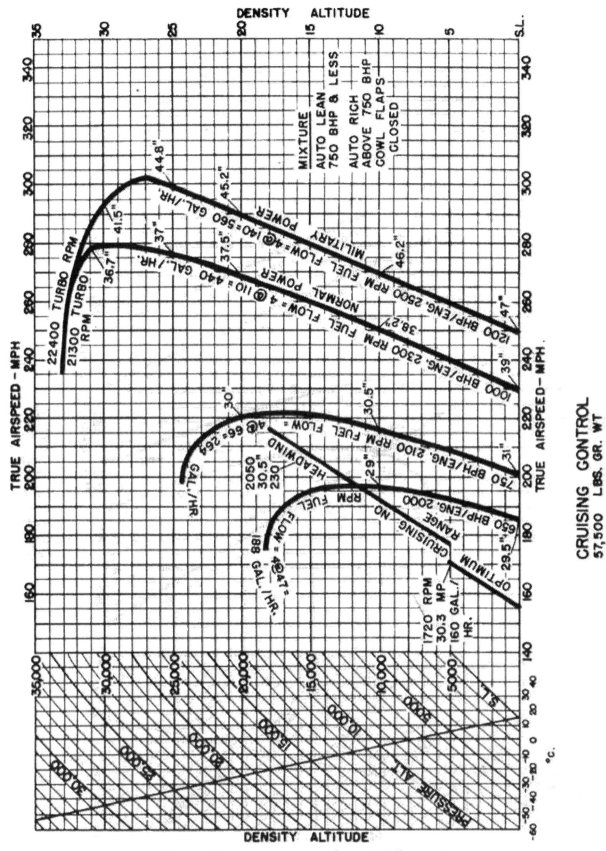

Cruising Control Chart - 57,500 Lbs.Gr.Wt.

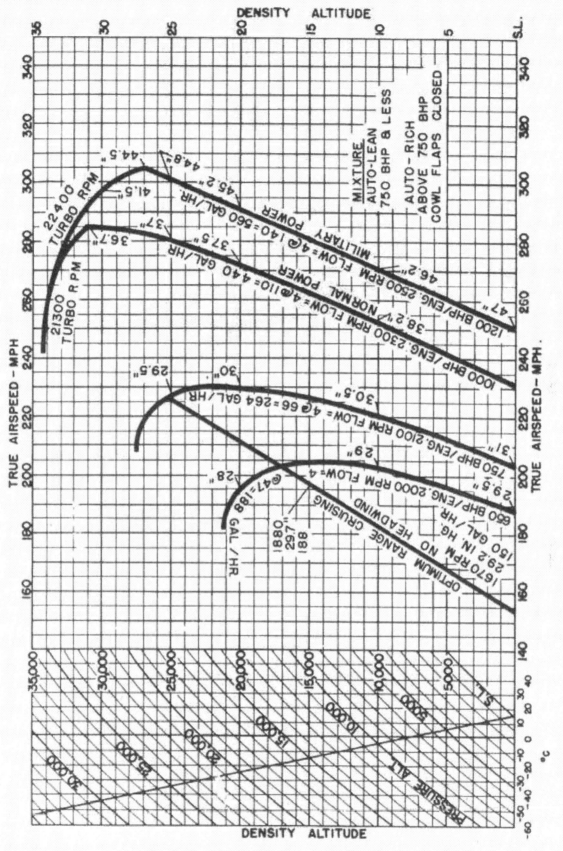

Cruising Control Chart - 55,000 Lbs. Gr. Wt.

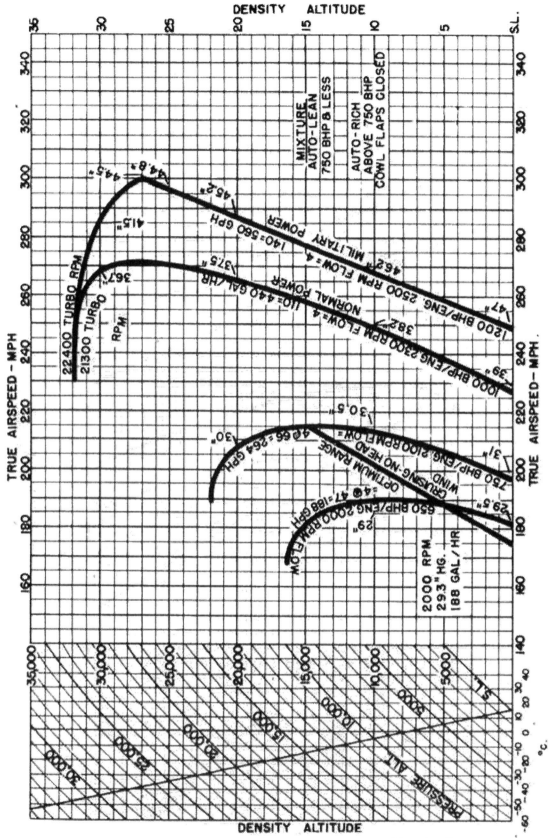

Cruising Control Chart - 60,000 Lbs.Gr.Wt.

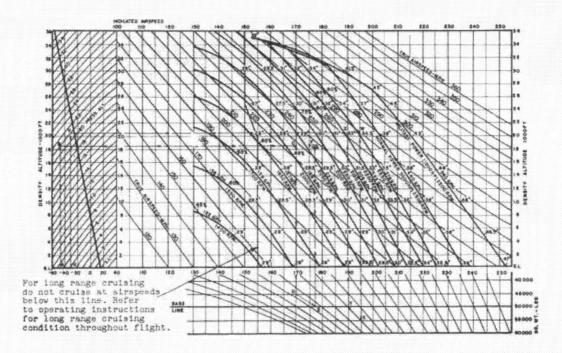

For long range cruising do not cruise at airspeeds below this line. Refer to operating instructions for long range cruising condition throughout flight.

This chart is suitable for use for all level flight conditions other than long range. Basically, this chart presents the performance of the airplane at 50,000 pounds gross weight at various RPM's in a plot of density altitude versus indicated airspeed. Diagonal lines running across the chart show the true airspeed as a function of indicated airspeed and density altitude. At the bottom of the chart there is a scale for converting the speed at 50,000 pounds to other gross weights.

For use in cruising flight, set manifold pressure and RPM to charted values as required to give speed or range desired. Determine density altitude, observed indicated airspeed. At charted manifold pressure and RPM in hot weather indicated airspeed will be low, in cold, high, when compared to charted values. Jockey power slightly as required (increase manifold pressure to increase speed, decrease manifold pressure to decrease speed) until charted IAS is obtained. Do not increase manifold pressures more than three inches above charted values without raising RPM. <u>Do not exceed 51 inches</u> manifold pressure or <u>2100 RPM</u> for continuous cruising in auto lean.

For steady cruising it should not be necessary to vary power oftener than every three hours.

Example: Follow arrows through A, B, C, D, E and F for gross weight less than 50,000 pounds to find true and indicated airspeed from temperature, pressure altitude, percent power required and gross weight. Point "C" is power required to make good speed at point "D" with weight of point "D".

For gross weights greater than 50,000 pounds follow arrows from U to Z.

Sample Problem: Total fuel is 1700 gallons, weight is approximately 55,000 pounds, bomb load 6,000 pounds, flight to be at 14,000 feet altitude. Objective is 400 miles distance or a total of 800 miles to be traveled. Determine power conditions for maximum speed to safely make round trip.

1. Assume 70 gallons for warm-up and takeoff; determine fuel to climb to 14,000 feet as 120 gallons and distance traveled in climb as 40 miles. Remaining fuel is 1510 gallons and distance is 760 miles.

2. Assume 310 gallons for headwinds, wandering and reserve. Fuel to be used is 1200 gallons.

3. At 14,000 feet altitude check speed at 2250 RPM and 378 gallons per hour. Speed at 55,000 pounds is 250 miles per hour or 200 miles per hour indicated airspeed.

4. Range is Fuel Available $\times \frac{\text{miles per hour}}{\text{gallons per hour}}$ which equals

$1200 \times \frac{250}{378} = 793$ miles.

Therefore, at 2250 RPM and 35 inches, 793 miles can be safely flown at a speed of 250 miles per hour or 200 miles per hour indicated airspeed at 14,000 feet altitude. With this condition 310 gallons will be left for reserve. Since fuel is used during the flight and the weight is decreased, this figure is conservative. If the bombs are dropped, the range above will be even more conservative. If the power conditions are not changed, the speed will increase by approximately 10 miles per hour from the start of the flight to the finish.

Composite Cruising Control Chart

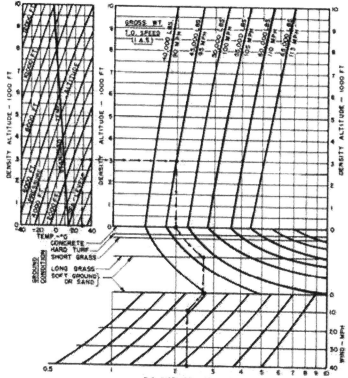

TAKEOFF DISTANCE OVER 50' OBSTACLE - ONE THIRD FLAPS — USE ABOVE TAKEOFF INSTRUCTIONS;
DISTANCE TO LEAVE GROUND IS APPROX. 85% OF THE TOTAL DISTANCE.
EXAMPLE: FOLLOW ARROWS AS INDICATED FOR CONDITION OF 30° AIR TEMPERATURE, 1000' ALTIMETER READING,
45,000 LBS. GROSS WEIGHT OF AIRPLANE, LONG GRASS RUNWAY, AND 10 MILE PER HOUR HEADWIND.
ANSWER: 2230' TO TAKE OFF AND CLIMB OVER 50' OBSTACLE.

Takeoff distances shown on the chart are optimum. Prior to takeoff run up engines at full throttle and set and lock turbos at 46 inches. For takeoff use 46 inches, 2500 RPM and 1/3 flaps and hold three-point position until airplane leaves the ground. Take off and hold at airspeed shown on chart until over obstacle. Distance to leave ground is approximately .85 times the total distance shown by chart.

Example: To determine takeoff distance for conditions of 30° outside air temperature, 1000 feet altimeter reading, 45,000 pounds gross weight of airplane, long grass runway and 10 MPH headwind.

1. Follow dash line from temperature of 30°C to 1000 feet pressure altitude to find density altitude of 3000 feet.

2. Project across 45,000 pounds and down to base line.

3. Follow curve to horizontal line representing long grass and go straight down to zero wind line.

4. Follow sloping wind lines to 10 MPH wind and go straight down to answer, 2230 feet, on bottom line.

5. In upper chart under gross weight note also indicated airspeed for takeoff and climb over obstacle.

Takeoff Control Chart

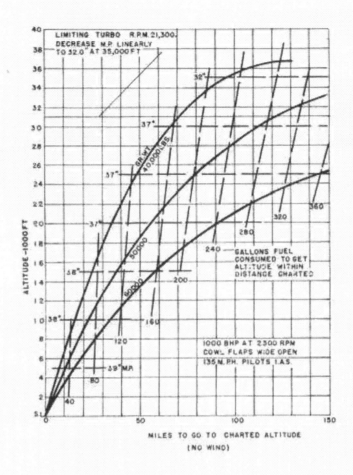

For best climbing conditions use 38", 2300 RPM, auto rich, cowl flaps open, and 135 MPH pilot's indicated airspeed. On instrument climbs below 20,000 feet, climb at 160 MPH pilot's indicated airspeed. Use full throttle and set power with turbo regulator. Decrease manifold pressure 1-1/2 inches per 1,000 feet above 30,000 feet. Climbs at lower RPM and manifold pressure will not result in improved range.

To determine distance travelled and fuel used in climbing to desired altitude refer to the Climb Control Chart.

Example: Gross weight is 60,000 pounds, desired altitude for flight is 20,000 feet. On vertical scale at left of chart locate 20,000 feet altitude line and follow to intersection of 60,000 pound line. On scale at bottom read 90 miles to charted altitude and on diagonal dashed line read 240 gallons of fuel consumed to get altitude within distance charted.

Note: Some improvement in climb performance can be obtained when the temperature is below standard by closing the cowl flaps down to the allowable cylinder temperature limits; however, this procedure is only recommended during combat.

Climb Control Chart

- 85 -

RESTRICTED

CRUISE CONTROL

FUEL CONSUMPTION

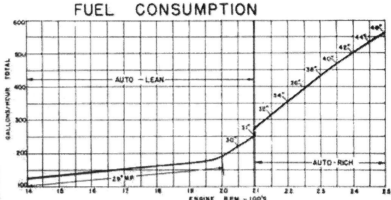

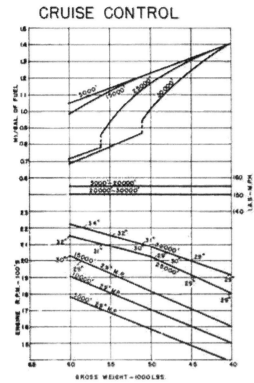

For long range cruising below 15,000 feet set RPM to maintain 155 MPH pilot's indicated airspeed with 29 ±1 inch manifold pressure. Above 20,000 feet use 150 MPH pilot's indicated airspeed and 29 inches. Use auto lean mixture. Close cowl flaps or set to obtain proper cylinder temperature. Hold power setting and let airspeed increase as fuel is used. Reset power every three hours to maintain desired cruising speed. The use of these instructions does not require a knowledge of the gross weight or RPM in order to set conditions inasmuch as long range cruising conditions are obtained automatically by adjusting to 155 MPH below 15,000 feet and 150 MPH above 15,000 feet at 29 inches manifold pressure.

To determine approximate RPM and miles per gallon at a given gross weight, refer to long range cruise control chart.

Example: Takeoff weight is 60,000 lbs; flight to be at 10,000 feet.

1. Allow 70 gallons for warm-up and takeoff and 100 gallons for climb to 10,000 feet.

    Resulting weight = 60,000# - 170 x 6 = 59,000# approximately

2. On cruise control chart at 10,000 feet at 59,000# read approximately 1900 RPM for 155 MPH pilot's indicated airspeed.

3. On fuel consumption chart at 1900 RPM and 29 inches read 170 gallons per hour fuel consumption.

4. Weight at end of three hours is

    59,000 - 170 x 3 x 6 = 56,000# approximately

    At 56,000# read approximately 1880 RPM and 1.13 miles per gallon.

5. Repeat above step every three hours.

The RPM's shown on this cruise control chart are affected by changes in configuration such as additional nose guns, opened windows, open cowl flaps, and to an extent by outside air temperature. It should therefore be remembered that the basic rule for long range cruising is 155 MPH and 29 inches below 20,000 feet and at the RPM should be adjusted accordingly for level flight.

Long Range Cruising Control Chart

RANGE VS. AIRSPEED

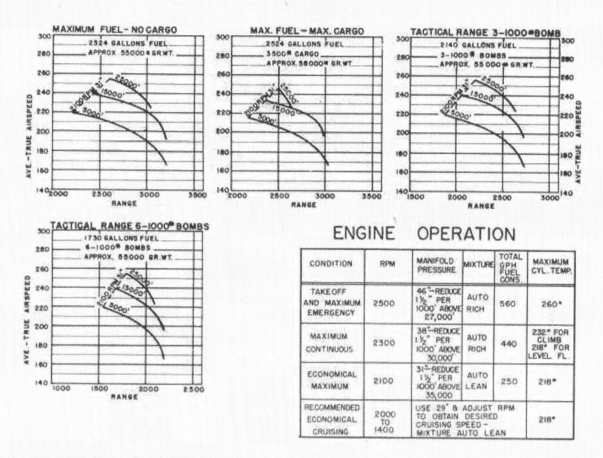

THE ABOVE CURVES ARE COMPUTED FROM INSTANTANEOUS CRUISING CONDITIONS OF ALTITUDE, POWER AND FUEL FLOW. REFER TO OPERATING INSTRUCTIONS FOR CRUISING CONDITIONS THROUGHOUT FLIGHT. NO CONSIDERATION IS MADE FOR HEADWINDS, WARM-UP, TAKEOFF, CLIMB, OR DESCENT. WHEN BOMBS ARE CARRIED, THE BOMB LOAD IS CONSIDERED TO BE CARRIED HALF THE DISTANCE OF THE FLIGHT. FOR FUEL CONSUMED IN CLIMBING REFER TO B-17F CLIMB CONTROL CHART.

These charts show the average airspeed vs. range for various altitudes and representative loadings. For the maximum ranges shown the specified long range cruising procedure should be employed. For the minimum ranges shown the airplane should be flown at 2100 RPM, 31 inches MP and auto lean mixture. For other range conditions it is sufficiently accurate to cruise at the average true airspeed shown and adjust the power conditions as per the engine operation chart to obtain the desired airspeed. Note that the curves are shown as average true airspeed and not indicated airspeed. As noted on the chart no allowances have been made for headwinds, warm-up, takeoff, climb or descent. In general, it can be assumed that these variables can be taken care of adequately by assuming the actual range, that can be safely flown and still leave sufficient reserve, is approximately 80 to 90% of the chart ranges depending to a great extent upon the existing headwinds.

Range Vs. Average Airspeed

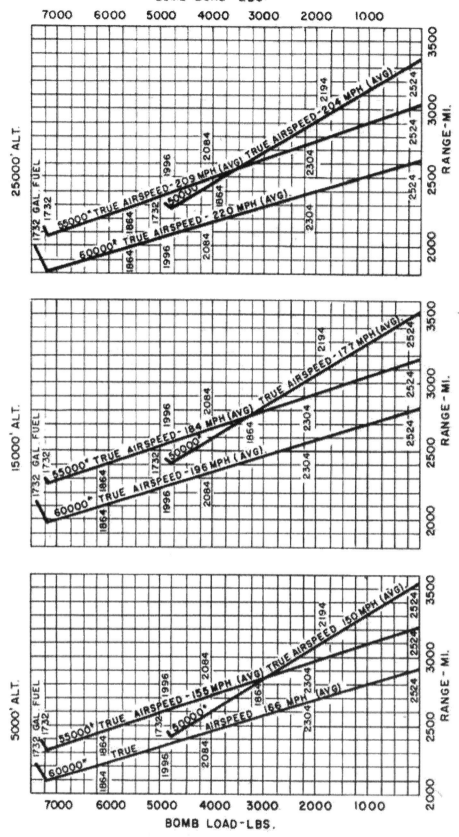

Range Vs. Bomb Load

- 88 -

RESTRICTED

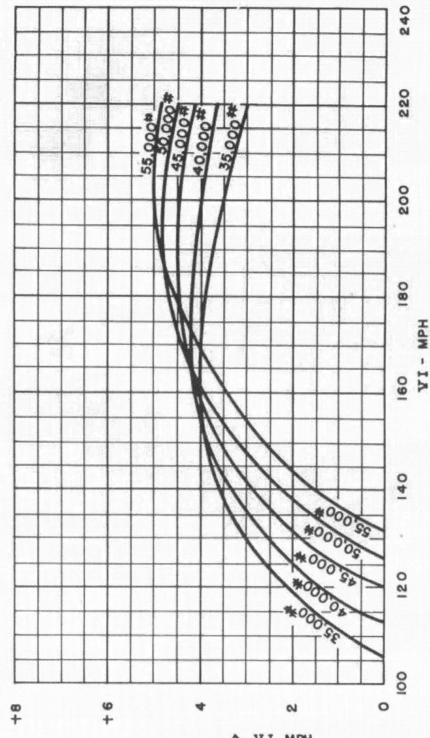

Airspeed Pitot Position Correction

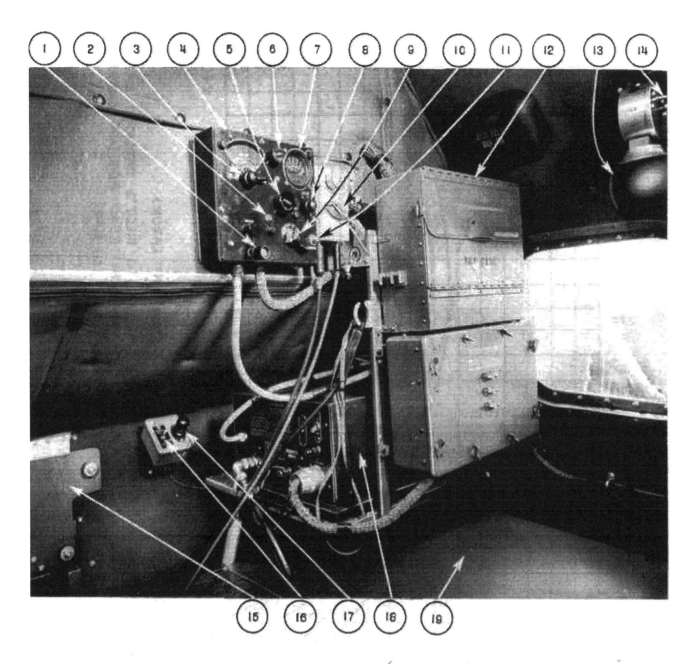

1 Tuning Crank	10 Interphone Jackbox
2 Control Indicator Lamp	11 Control Push Button
3 Band Selector Switch	12 Map Case
4 Radio Compass Control Unit	13 Navigator's Light
5 Volumn Control	14 Navigator's Light Switch
6 Light Control Switch	15 Armor Plate
7 Tuning Meter	16 Panel Light Switch
8 Loop Control Switch	17 Panel Light
9 Radio Compass Power Switch	18 Radio Compass Receiver
	19 Navigator's Table

Figure 46 - Navigator's Communication Controls

Figure 47
Navigator's Seat
Adjustment Diagram ➡

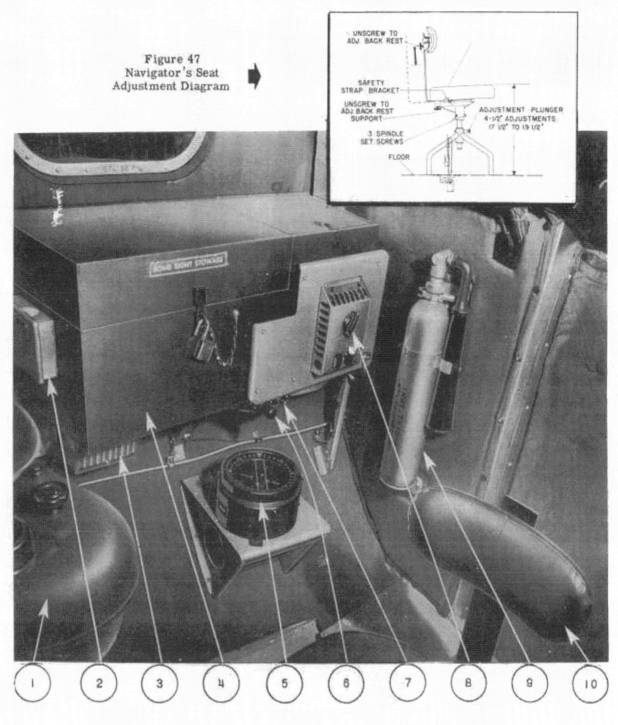

1 Drift Meter	5 Aperiodic Compass
2 Fuse Box	6 Panel Light
3 Heating and Ventilating System Opening	7 Panel Light Switch
	8 Suit Heater Outlet
4 Bomb Sight Stowage Box	9 Fire Extinguisher
	10 Navigator's Seat

Figure 48 – Navigator's Compartment – Right Rear Corner

1 Navigator's Table
2 Drift Meter Master Switch
3 Radio Compass Indicator
4 Ash Receiver
5 Drift Meter

Figure 49 - Navigation Equipment - Navigator's Compartment

SECTION IV

OPERATING INSTRUCTIONS - NAVIGATOR'S COMPARTMENT

1. **General Description.**

    a. The navigator's compartment is located in the nose of the airplane. It extends from the rear of the bombardier's seat to the bulkhead which forms the forward end of the pilot's compartment. The compartment contains oxygen control, interphone controls, necessary lights and controls, navigational equipment, the bomb sight storage box, a fire extinguisher, and controls for the radio compass.

    b. The radio compass receiver is supported above the aft end of the navigator's table and may be remotely controlled either from the pilot's compartment ceiling, or from the control unit on the receiver support on the navigator's table.

2. **Operational Equipment.**

    a. Initial Entrance. - The navigator's compartment is entered through a door in the bulkhead in the aft end of the compartment, which is reached by entering the main entrance door of the airplane, passing through the pilot's compartment through the small trap door between the pilot's and copilot's seats, or by entering the front entrance door (figure 120) in the bottom of the airplane below the pilot's compartment. The front entrance door is the escape hatch for the navigator's compartment.

    b. Lighting.

    (1) A dome light is located in the ceiling of the compartment, with a switch located adjacent to the light.

    (2) A panel light (figure 46-) and switch (figure 46-) are located above the navigator's table on the aft wall of the compartment.

    (3) The navigator's light (figure 46-) is located on the wall near the ceiling directly over the navigator's table. The switch (figure 46-) is located on the base of the lamp.

    c. Fire Extinguisher. - A hand CO_2 fire extinguisher (figure 48-) is clipped to the aft wall of the compartment to the right of the door.

    d. Seat Adjustment. - The navigator's seat is strapped to the floor. For complete adjustment instructions, see figure 47.

    e. Interphone Equipment. - The interphone jackbox (figure 46-) is located between the radio compass control box (figure 46-) and the map case (figure 46-). Operation is outlined in paragraph 4.j. of section I.

    f. Oxygen Equipment. - The oxygen regulator is located on the wall above the navigator's table. Operation is conventional.

    g. Heating and Ventilating System Inlet. - The inlet (figure 48-), located beneath the bomb sight storage box, is equipped with a push-pull knob for regulating the flow of air. Push to open and pull to close. The selection of hot or cold air is made by the pilot.

    h. Drift Meter Master Switch. - A master switch (figure 49-) for the drift meter (figure 49-) is located below the edge of the navigator's table near the ash receiver (figure 49-) on the front forward corner. Operation is conventional.

    i. Radio Compass Receiver. - Operation of the radio compass receiver (figure 46-) is the same for the navigator as it is for the pilot. Refer to paragraph 4.i., section I for detailed use of the radio compass receiver.

    j. Bearing Indicator. - The bearing indicator is mounted beneath the forward inboard corner of the navigator's table and its dial (figure 49-) may be seen by lifting the cover on the table. The loop antenna is remotely controlled from the radio compass receiver.

    k. Aperiodic Compass. - The compass is located on the right side of the compartment below the bomb sight storage box.

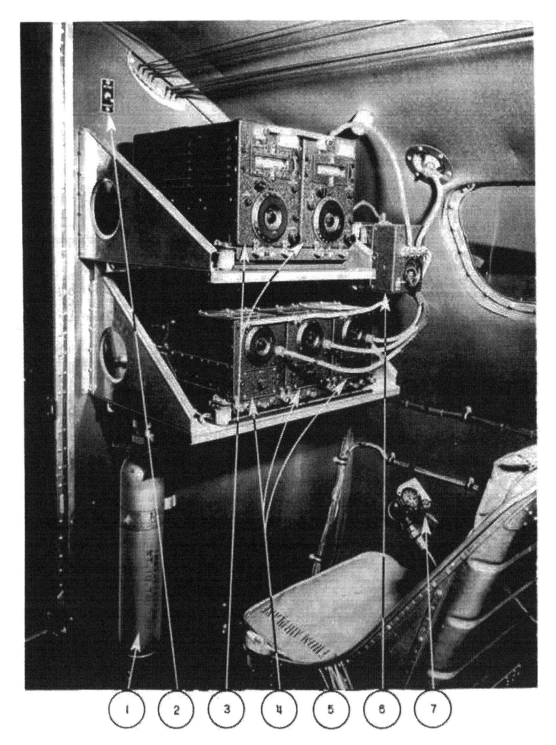

1. Fire Extinguisher
2. Bomb Bay Step Light Switch
3. Command Transmitters
4. Command Receivers
5. Seat For Auxiliary Crew
6. Antenna Relay Control Box
7. Oxygen Regulator

Figure 50 - Command Radio Installation - Radio Compartment

SECTION V

OPERATING INSTRUCTIONS - RADIO COMPARTMENT

1. General Description.

   a. The radio compartment is located immediately aft of the bomb bay. It contains most of the radio equipment, seats for the radio operator and two crew members, oxygen controls, interphone controls, an alarm bell, heating and ventilating system duct inlet, life raft controls, two emergency hand cranks with extensions, and a fire extinguisher.

   b. Two command radio transmitters and three receivers are mounted on the right side of the compartment on the forward bulkhead. (See figure 50.) They are controlled by remote control units on the ceiling of the pilot's compartment. The transmitters' dynamotor and modulator are on the floor in the forward right-hand corner of the compartment. The receivers' dynamotors are mounted on supports behind the receivers.

   c. The communications equipment consists of the following:

Command Set	SCR-274-N
Liaison Set	SCR-287-A
Radio Compass Recorder	SCR-269-C
Interphone Equipment	RC-36
Marker Beacon Equipment	RC-43
Radio Altimeter	SCR-518-A
IFF Radio Set	SCR-535-A

   d. The liaison transmitter is installed on the left-hand side of the aft bulkhead. The receiver is located on the radio operator's table. The dynamotor is located on the left rear side of the aft bulkhead in the ball turret compartment. Two antennae are available for use with the liaison set. One employs the skin of the airplane, with the lead-in attached to the change-over switch (figure 52-2) on the left side wall. The other is the trailing antenna which is also attached to the change-over switch. The trailing antenna reel is electrically operated from a control box (figure 52-4) to the right of the change-over switch.

2. Operational Equipment.

   a. Lighting.

   (1) A light (figure 56-5) is located above the radio operator's table with a switch located adjacent to the light. A similar light (figure 51-3) and switch (figure 51-2) are located in the aft end of the compartment above the liaison transmitter.

   (2) The bomb bay step light switch (figure 50-2) is located on the forward wall of the compartment above and to the left of the command transmitters.

   (3) A light (figure 58-3) and switch (figure 58-2) are located on the side wall to the left of the radio operator's seat.

   b. Emergency Equipment.

   (1) A fire extinguisher (figure 50-) is located on the forward wall of the compartment to the right of the door.

   (2) Two life raft control handles (figure 122-2 and -3) are located on the ceiling of the compartment just aft of the top hatch on the right-hand side. To release the life rafts, pull the handles approximately 9 inches.

   (3) Four red canopy (top hatch) emergency release handles are located along the edge of the canopy.

   (4) An alarm bell (figure 56-9) is located on the forward wall of the compartment above the radio operator's table.

   (5) Two hand cranks (figure 53-6) for manual operation of the wing flaps, bomb bay doors, landing gear, tail gear and starters, and two extensions, are clipped to the aft wall of the compartment above the transmitter tuning units. Detailed operating instructions for manual operation of the bomb bay doors and landing gear will be found in sections X and XIV, and for manual operation of the tail gear in sections XII and XIV. For manual operation of the wing flaps see the following paragraph and also section XIII.

   (6) For emergency operation of the wing flaps, open the camera pit door and insert a crank with extension attached into the connection at the forward end of the pit. Rotate the crank clockwise to lower the flaps and counterclockwise to raise them.

   (7) Portable Emergency Transmitter. - Complete operating instructions will be found on the face of the transmitter itself. A complete Handbook of instructions will be found inside the transmitter bag. (See figure 53-2.)

   c. Seat Adjustment. - For complete instructions for adjusting the seats, see figure 54.

   d. Interphone Controls. - Detailed operating instructions for the use of the interphone system will be found in section I, paragraph 4.j. The radio operator's interphone jackbox (figure 56-2) is located on the left side wall. The jackbox for the other crew members (figure 55) is located to the right side wall of the compartment.

e. Oxygen Controls. - Oxygen outlets exist for the radio operator (figure 56-1) and for each of the two auxiliary crew members. (See figures 50-7 and 55.) Operation is conventional.

f. Heating and Ventilating System Inlet. - The inlet (figure 52-5) is located on the floor of the compartment to the left and aft of the radio operator's seat. Push the knob to close and pull to open.

g. Suit Heater Outlets. - The heat output of the suit is controlled by a rheostat on the two receptacle boxes. (See figures 52-4 and 57.)

h. Radio Set, SCR-518-A (High Altitude Altimeter).

(1) Description. - Radio set SCR-518-A consists of a complete set of apparatus for determining the height of the airplane above the ground. It is operative over an altitude range of 0-20,000 feet, but it will work satisfactorily up to 30,000 feet before the indications become erroneous. Operation of the set does not depend upon barometric pressure. It indicates altitude of the aircraft above the terrain below the airplane, and has no reference to sea level. If the aircraft is flying over broken country, more than one peak will appear on the indicator, the highest one representing the object closest to the airplane.

(2) Operation.

(a) Place the power on-off switch in the "ON" position. This energizes all parts of the set except the automatic volume control which is controlled by a separate switch. A pilot lamp located at the lower center of the control panel should light, indicating that the power is on.

(b) As the tubes reach their operating conditions, the circle traces and indicating lobes appear on the screen of the indicator. During the first few minutes of operations the indications will be unsteady.

(c) Turn the "CIRCLE SIZE" control knob until the two circle traces on the indicator screen are adjusted to the required diameter for readings. The proper size occurs when each circle is just visible as a luminous green ring on the gray background just beyond the outer circumference of its dark calibrated scale ring.

(d) Turn the "RECEIVER GAIN" control to adjust the lobe readings for clearest legibility on the indicator screen. Maximum receiver sensitivity may be used at the higher altitudes and less than maximum sensitivity may be required at the lower altitudes. The receiver gain control must be adjusted in conjunction with the automatic volume control switch for maximum lobe legibility on the altimeter scale in accordance with the following paragraphs.

1 Liaison Antenna Tuning Unit
2 Light Switch
3 Light
4 Liaison Transmitter
5 Transmitter Tuning Unit

Figure 51 - Liaison Radio Transmitter Installation

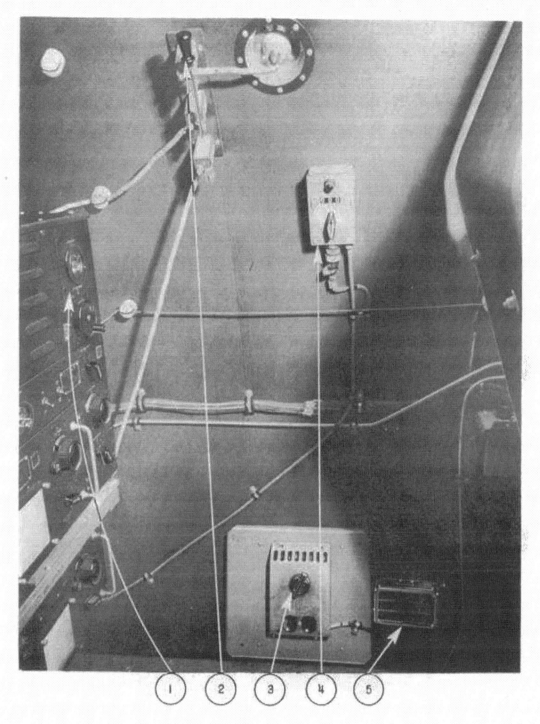

1 Liaison Transmitter
2 Antenna Change-Over Switch
3 Suit Heater Outlet
4 Trailing Antenna Reel Control
5 Heating & Ventilating System Inlet

Figure 52 - Radio Compartment - Left-Hand Side

1. Seat For Auxiliary Crew
2. Portable Emergency Transmitter
3. Transmitter Tuning Units
4. Starter Crank Extension
5. Extension Of Bomb Doors And Flaps
6. Hand Crank
7. Door To Ball Turret Compartment

Figure 53 - Radio Compartment - Right Side - Looking Aft

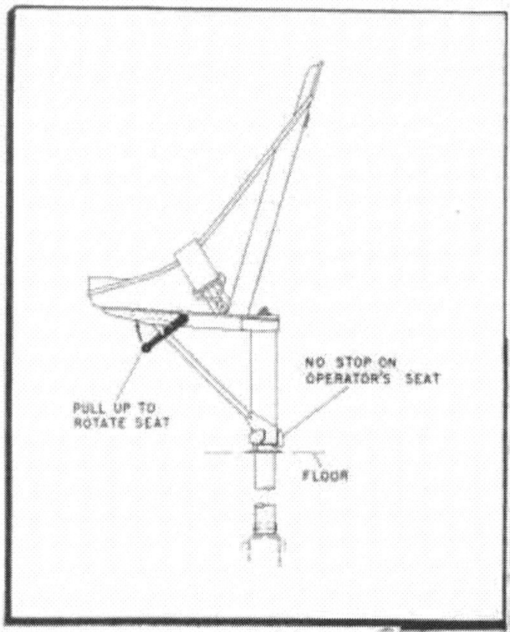

Figure 54

Radio Operator's Seat Adjustment Diagram

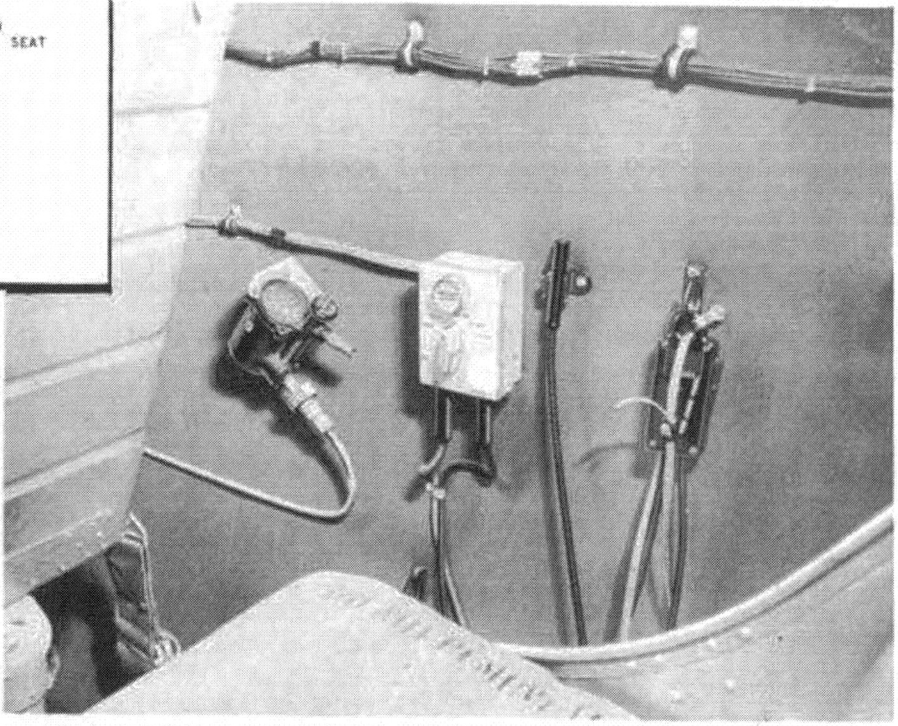

Figure 55

Auxiliary Crew's Interphone Controls

(e) <u>Use of Automatic Volume Control at Lower Altitudes</u>.

1. The automatic volume control improves the performance of the radio set at altitudes below 2000 feet and should only be used for reading up to 2000 feet. With the AVC switch on, receiver sensitivity is reduced but is automatically increased with altitude up to about 2000 feet. Overloading of the receiver is thus prevented at the lower altitudes.

2. For operation when descending below 2000 feet:

a. At any altitude above 1000 feet, throw AVC switch on.

b. Adjust "RECEIVER GAIN" control until the initial lobe appearing at zero on the 2000-foot scale is the proper height.

c. The reception lobe giving the altitude reading on the 2000-foot scale should now remain approximately constant in size as the ground is approached.

(f) <u>Use of AVC at Higher Altitudes</u>. - The AVC switch must be turned off when the equipment is operating altitudes above 2000 feet, as the AVC would otherwise impair the receiver sensitivity in certain sections of the higher altitude ranges.

(g) Starting from zero and reading in a clockwise direction, read the <u>counterclockwise</u> edge of each lobe on each circle trace. (If the lobe is on the top of the dial, read to the left edge, and if it is at the bottom of the dial, read the right edge.) The first lobe (or index lobe) appears at the zero calibration on each scale. The second lobe (reflection lobe) indicates the altitude above terrain.

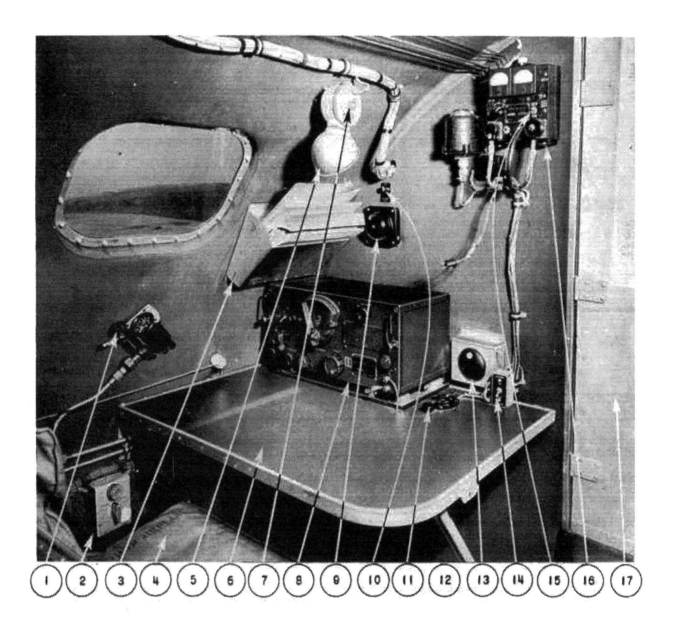

1 Oxygen Regulator
2 Interphone Jackbox
3 Report Rack
4 Radio Operator's Seat
5 Radio Operator's Light
6 Radio Operator's Table
7 Light Switch
8 Liaison Set Receiver
9 Alarm Bell
10 Phone Call Lamp
11 Transmitting Key
12 Radio Set SCR-535 Crash Switch
13 Ash Reciver
14 Liaison Transmitter Master Switch
15 Local "Off-On" Switch SCR-535
16 Radio Set SCR-535 Control Box
17 Door To Bomb Bay

Figure 56 - Radio Operator's Table and Controls

Figure 57 - Suit Heater Outlet Right Side Radio Compartment

1. On each scale (inner and outer), the index lobe will appear at the zero calibration. The second (reflection lobe) on each scale indicates the absolute altitude of the aircraft.

2. The inner circle is merely a vernier on the outer circle. On the outer circle, it is possible to read to within 250 feet. If greater accuracy is required, the inner scale reading must be taken into consideration, as follows: Read the outer scale to the next lower even thousand (4000, for instance). Read the inner scale. If the reading of the inner scale should be 750 feet, the actual altitude of the aircraft is then obtained by adding the readings of the two scales: 4750 feet. The inner scale can, with practice, be read to within 25 feet.

3. If the zero lobes have shifted away from zero, correct readings may be obtained by adding the amount of zero shift if the shift is to the left of zero, and by subtracting the amount of zero shift if the shift is to the right, from the reading of altitude which was obtained by following the procedure outlined in the preceeding paragraph.

4. If the altitude of the aircraft exceeds 20,000 feet, the indications will still indicate the absolute altitude, providing the operator adds exactly 20,000 feet to the apparent altitude. It will usually be obvious (from the pressure altimeter, for instance) when the aircraft is at or above 20,000 feet. If the operator knows that the aircraft is at least 20,000 feet high, but the indicator shows only 2700 feet, the actual altitude is obviously 22,700 feet.

5. Flying over rough terrain will show fluctuating indications and flying over water will give a steady indication.

(3) Precautions.

(a) Under normal variable service conditions, the following inaccuracies of performance are to be expected:

1. Initial deviation from perfect accuracy 50 feet plus 1/4 of 1 percent of the time or absolute altitude.

2. At zero altitude, the maximum error in reading will be 50 feet.

3. At 20,000 feet, the maximum error in reading will be 100 feet.

EXAMPLE:

Maximum reading at 20,000 feet:
$20,000 + 50 + (.0025 \times 20,000) = 20,100$ feet
Minimum reading at 20,000 feet:
$20,000 - (50 + .0025 \times 20,000) = 19,900$ feet

(4) Observable Defects.

(a) Circle traces of unsymmetrical shape will cause inaccurate readings.

(b) Circle traces badly off center will cause inaccurate readings.

(c) Shifting of zero transmission lobes will cause incorrect scale readings, but correct readings may be obtained by adding or subtracting the shift reading of the zero lobe to or from the indicator reception lobe reading.

h. Armor Protection. - Armor sufficient to withstand U.S. .30, German .312, and Japanese and Italian .303 (7.7 mm) caliber fire by direct right angle hit is provided for the radio operator. Enemy fire originating within the areas illustrated in figure 59 will not reach the pilot.

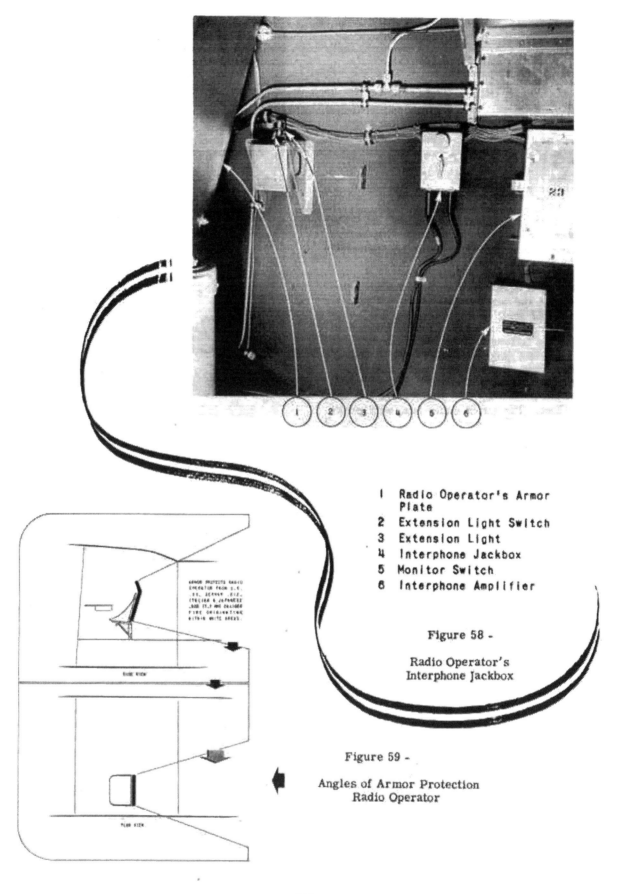

1. Radio Operator's Armor Plate
2. Extension Light Switch
3. Extension Light
4. Interphone Jackbox
5. Monitor Switch
6. Interphone Amplifier

Figure 58 -

Radio Operator's Interphone Jackbox

Figure 59 -

Angles of Armor Protection Radio Operator

SECTION VI

OPERATING INSTRUCTIONS - BOMBARDIER'S COMPARTMENT

1. General Description.

The bombardier's compartment is located in the nose of the airplane. There is no physical separation between it and the navigator's compartment. The compartment contains bomb controls, instrument and control panels, bomb rack indicator (see inset), interphone jackbox, oxygen control, machine gun, heating and ventilating system inlet, bomb sight window defroster and control, and a heated suit outlet.

2. Operational Equipment.

a. Lighting.

(1) No dome light is located in this compartment. Light is received from the dome light in the navigator's compartment.

(2) A flexible light (figure 62-8) and switch rheostat (figure 62-7) are located forward and above the bomb controls on the left side of the compartment.

(3) An extension light (figure 63-11) and switch (figure 63-10) are located on the bombardier's instrument panel.

b. Emergency Equipment. - The emergency escape route is through the navigator's compartment and out the front entrance door in the bottom of the fuselage below the pilot's compartment.

c. Seat Adjustment. - See figure 60 for complete instructions for adjustment of the seat.

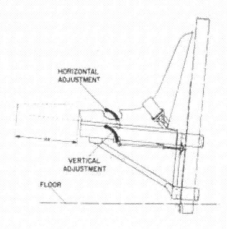

Figure 60 - Bombardier's Seat Adjustment Diagram

d. Interphone Controls. - An interphone jackbox is located on the right side of the compartment. Detailed operating instructions are outlined in section I, paragraph 4.j. (See figure 61-1.)

e. Bomb Controls.

(1) Bombs are normally released electrically, but can be mechanically released in an emergency. Electrical control provides for individual release of bombs either singly (selective) or continuously at predetermined intervals (train). Mechanical control is always in "SALVO" by operation of the bomber's release handle or by operation of the emergency release handle. The bomb release handle (figure 62-5) has three positions as follows:

(a) Lock. - In the "LOCK" position the bomb racks are locked against any release of bombs except by means of the emergency release handles.

(b) Selective. - In the "SELECTIVE" position the bomb racks are prepared for electrical release by manual operation of the release switch or by automatic operation through the bomb sight.

(c) Salvo. - Replacement in the "SALVO" position when the bomb doors are open, mechanically releases all bombs simultaneously and unarmed.

(2) The bombardier's release switch (see inset) is mounted on the forward end of the control panel. This is a double throw momentary contact toggle switch which operates in either direction to energize the release unit solenoids through the interval release control mechanism. A hinged guard (figure 62-1) prevents inadvertent operation of this switch.

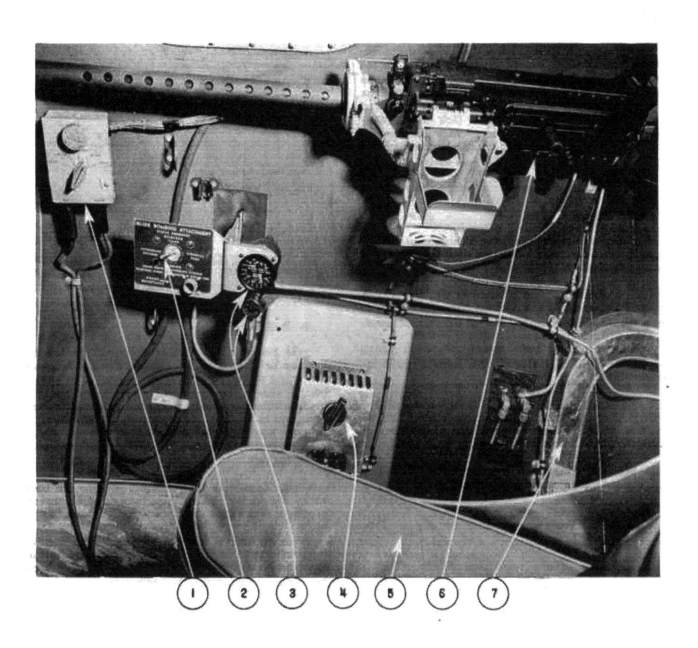

1 Interphone Jackbox
2 Glide Bombing Attachment Static Pressure Selector Valve
3 Oxygen Regulator
4 Suit Heater Outlet
5 Bombardier's Seat
6 .30 Caliber Machine Gun (Stowed)
7 Heating & Ventilating System Duct

Figure 61 - Bombardier's Controls - Right Side Wall

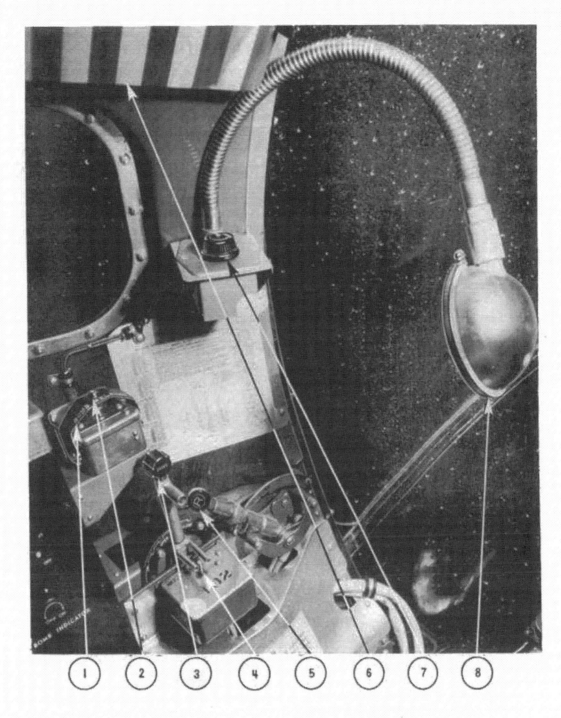

1. Bomb Release Switch Guard
2. Bomb Release Switch
3. Bomb Door Control Handle
4. Bomb Door Switch
5. Bomb Release Handle
6. Clipboard
7. Bombardier's Light Switch
8. Bombardier's Light

Figure 62 - Bombardier's Bomb Controls

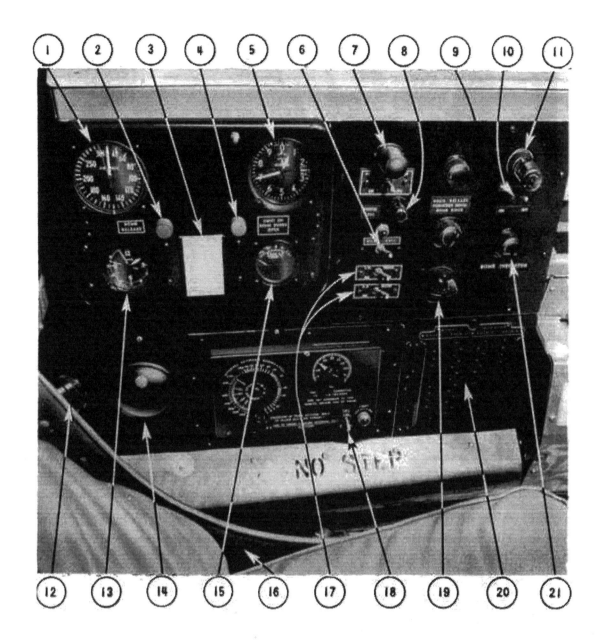

1	Air Speed Indicator	12	Ultra-Violet Spot Light
2	Bomb Release Warning Lamp	13	Clock
3	Altimeter Scale Error Card	14	Ash Receiver
4	Bomb Door Warning Lamp	15	Free Air Thermometer
5	Altimeter	16	Bombardier's Seat
6	Pilot Call Switch	17	Bomb Rack Selector Switches
7	Panel Light	18	Bomb Interval Switch
8	Phone Call Lamp	19	Ultra-Violet Spotlight Control Switch
9	Warning Lamp Rheostat	20	Bomb Indicator
10	Extension Light Switch	21	Bomb Indicator Control Knob
11	Extension Light		

Figure 63 - Bombardier's Instrument Panel

MAXIMUM AIRPLANE GLIDE & CLIMB ANGLES FOR BOMB RELEASE

WITH WHEELS AND FLAPS UP: MAXIMUM ALLOWABLE INDICATED AIR SPEED IS 305 M.P.H. SAFE GLIDE ANGLE IS 15 1/4°.

WITH WHEELS AND FLAPS DOWN: MAXIMUM ALLOWABLE INDICATED AIR SPEED IS 147 M.P.H. SAFE GLIDE ANGLE IS 13 1/2°.

NOTE: THE SAFE GLIDE ANGLES ARE BASED ON AN AIRPLANE GROSS WEIGHT OF 40,000 LBS. WITH POWER OFF AND WINDMILLING PROPELLERS.

WHILE THE MAJORITY OF BOMB STATIONS WILL PERMIT RELEASE OF BOMBS AT AN ANGLE WHICH WILL PRODUCE AN INDICATED AIR SPEED GREATER THAN THAT DESIGNATED FOR THE SAFE GLIDE ANGLE OF THE AIRPLANE, UNDER NO CONDITIONS SHALL THE MAXIMUM ALLOWABLE INDICATED AIR SPEED BE EXCEEDED.

ANGLES SHOWN ALLOW 10° FOR SAFETY. HOWEVER, UNDER PERFECTLY SMOOTH FLYING CONDITIONS, IF IN THE AIRPLANE COMMANDER'S OPINION CONDITIONS WARRANT IT, THESE GIVEN ANGLES MAY BE EXCEEDED BY NOT MORE THAN 5°.

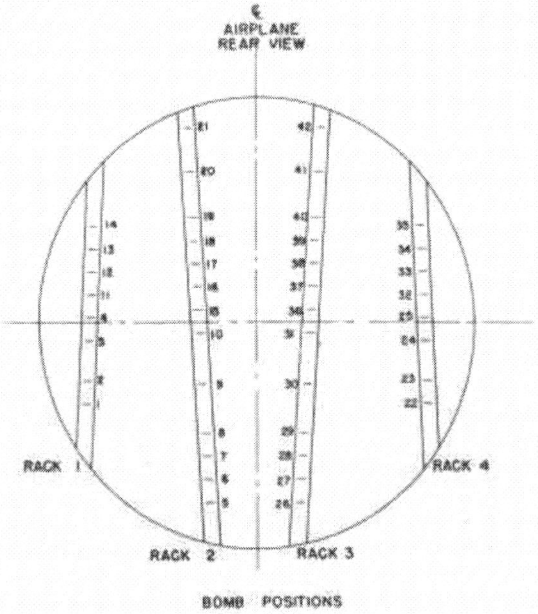

THE GLIDE OR CLIMB ANGLE IS THE ANGLE INCLUDED BETWEEN THE EARTH'S SURFACE AND THE FUSELAGE CENTERLINE.

THE ANGLES LISTED IN THE TABULATION ARE THE MAXIMUM AT WHICH BOMBS MAY BE RELEASED WITH A 10° CLEARANCE ANGLE MAINTAINED IN THE BOMB BAY.

Figure 64 - Bomb Release Angles Chart

RESTRICTED

(3) The interval release control unit is mounted at the bottom of the bombardier's control panel and may be set to provide either "SELECT" or "TRAIN" release. On airplanes serial Nos. 42-5050 and on, four switches on the bombardier's control panel permit selection of any external or internal rack for electrical release. Two additional rack selector switches in the bomb bay permit elimination of either right-or left-hand bomb bay from the release circuit if bomb bay fuel tanks are carried. The bomb release sequenced with all six rack selector switches "ON" is:

1 - LH internal station No. 1
2 - RH external rack
3 - RH internal station No. 22
4 - LH external rack
5 - LH internal station No. 2
6 - RH internal station No. 23

and continuing alternately between left-hand and right-hand internal bays. Any rack or combination of racks may be eliminated from the release sequency by turning off the respective selector switch on the bombardier's control panel.

(4) A bomb door control handle is located to the left of the bombardier forward of the control panel, and operates a double throw toggle switch which controls the solenoid switches for the bomb door retracting motor. A lug on the side of the handle is located so that when the door handle is in the "CLOSED" position, the bomb release lever cannot be moved out of the "LOCK" position.

CAUTION: If bombs are carried above the 2000-pound bomb, they MUST NOT be released until the D-6 shackle and adapter have been removed. This definitely requires "SELECTIVE" release control for the 2000-pound bomb.

f. Oxygen Control. - The regulator (figure 61-3) is located on the right-hand wall of the compartment. Operation is conventional.

g. Heating and Ventilating System. - The duct has a control knob which controls the flow of air to the bomb sight window located in the floor in front of the bombardier's seat. The knob is shoved forward to shut off the flow of air and pulled aft to allow air to reach the bomb sight window. Selection of hot and cold air is made by the pilot.

Figure 65 - Bombsight Window Defroster Control

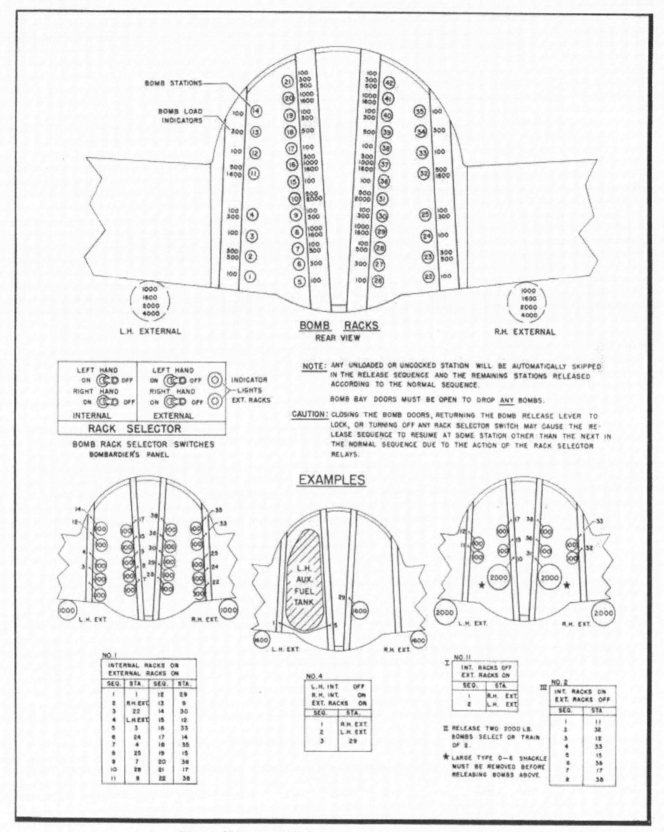

Figure 65A - Bomb Release Sequence Diagram - Sheet 1

RESTRICTED T. O. No. 01-20EF-1

ANY BOMB LOAD WILL BE RELEASED ACCORDING TO ONE OF THESE SEQUENCES. COMBINATIONS OF RELEASE SEQUENCES FOR A PARTICULAR BOMB LOAD ARE POSSIBLE BY OPERATION OF THE RACK SELECTOR SWITCHES BETWEEN "STICKS." (SEE CAUTION ON SHEET NO.1)

Figure 65B - Bomb Release Sequence Diagram - Sheet 2

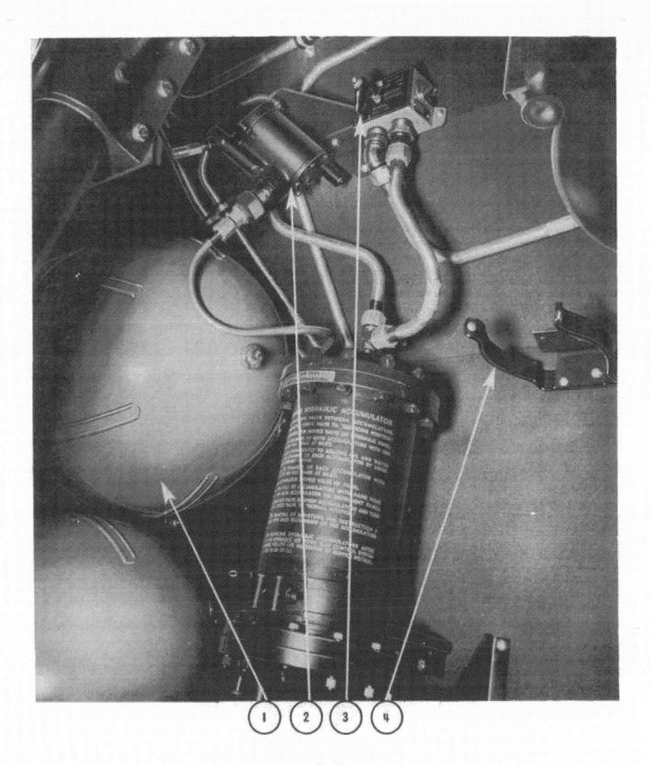

1 Oxygen Bottle
2 Oxygen Regulator
3 Selective Check Valve
4 Thermos Bracket

Figure 66 - Selective Check Valve

RESTRICTED

1 Gun Sight
2 Sight Light Rheostat Control
3 Sight Switch
4 Gun Charging Handles

Figure 67 - Upper Turret-Interior View

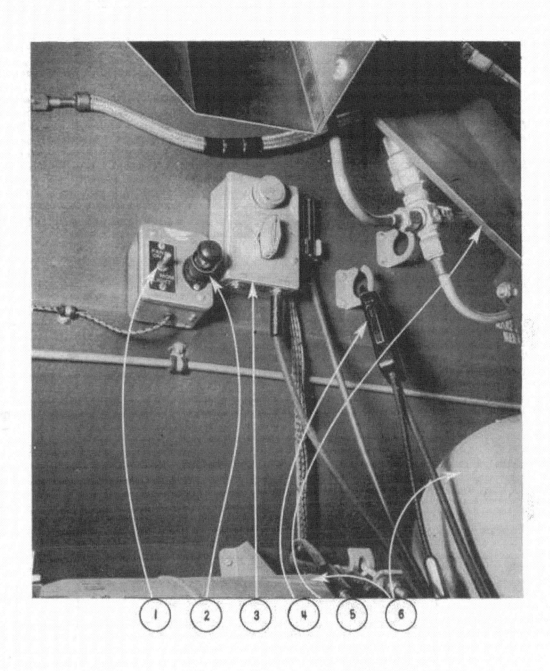

1. Panel Light Switch
2. Panel Light
3. Interphone Jackbox
4. Throat Microphone Push-to-talk Switch
5. Turret Structure
6. Oxygen Bottle

Figure 68 - Upper Turret Operator's Interphone Jackbox

SECTION VII

OPERATING INSTRUCTIONS - UPPER TURRET COMPARTMENT

1. General Description.

The upper turret compartment is located directly behind the pilots' seats. It is equipped with oxygen and interphone controls, two thermos bottles, fire extinguisher, oxygen bottle for the turret operator, the two fuel transfer valves and pump switch, eight oxygen bottles which furnish part of the oxygen for the airplane's master system, the upper gun turret containing twin .50 caliber machine guns, and the selective check valve for the hydraulic system.

2. Operational Equipment.

a. Lighting.

(1) Light for the compartment in general is furnished from the dome light (figure 4-2) on the ceiling between the pilot's and copilot's seats. The switch (figure 4-1) is located directly aft of the light.

(2) A panel light (figure 68-2) and switch (figure 68-1) are located on the wall of the compartment to the left of the turret.

(3) A trouble light (figure 72-7) and switch (figure 72-8) are located on the inside of the turret on the right-hand side looking aft.

b. Interphone Controls. - An interphone jackbox (figure 68-3) is located on the wall of the compartment to the left of the turret next to the panel light. Detailed operating instructions are given in section I, paragraph 4.j.

c. Oxygen Controls. - One oxygen regulator (figure 66-2) is located on the wall of the compartment to the right of the turret beside the selective check valve. A second regulator (figure 73-6) is located inside the turret on the right side looking aft. This regulator is connected with the bottle attached to the lower part of the turret. When the supply of this bottle is exhausted, it can be refilled from the airplane's master system by using the connection provided on the outside of the turret. Operation of the regulator is conventional.

d. Armor Protection. - The aerial engineer (turret operator) is protected from enemy fire by armor plate on the bulkhead between the turret compartment and the bomb bay. See figure 69 for angles of protection.

e. Fuel Transfer Controls.

(1) Selection of source and destination for fuel transfer is made by means of two fuel transfer valves (figure 70-2) located below the door between the upper turret compartment and the bomb bay on the forward side. Each valve has four ports. One port is connected to a port of the transfer pumps and will be the inlet OR outlet depending upon the direction of transfer. The other three valve ports are connected to the two engine tanks and the bomb bay tank (when installed) on their respective sides of the airplane. The transfer of fuel must always cross the center line of the airplane. In order to transfer between two tanks on the same side of the airplane, it is necessary first to transfer to a tank on the opposite side of the airplane, then reset one valve for the final destination and transfer back to the desired tank. Refer to the fuel transfer operation diagram (figure 71) for additional information and instructions.

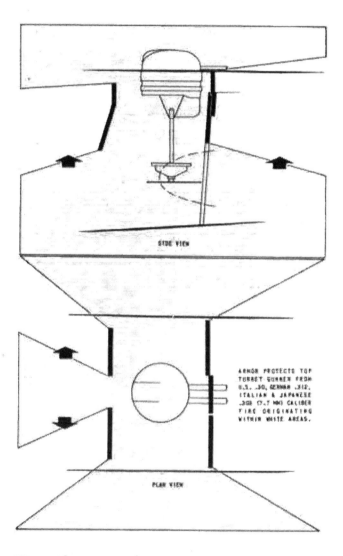

Figure 69 - Angles of Armor Protection - Top Turret

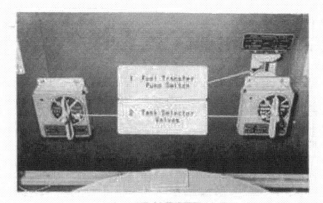

Figure 70 - Fuel Transfer Controls - Looking Aft

(2) The switch (figure 70-1) for the fuel transfer pump is composed of two double throw toggle switches with the handles linked together. The direction of throw of the switch handle corresponds to the direction of flow of fuel. In order to prevent operation of the transfer pump against a closed valve, a safety switch at each valve holds the electrical circuit open except when the valve is fully turned to one of the three tank positions.

WARNING: Do not inadvertently use the bomb bay valve position when the bomb bay tanks are not installed. No provision is made for elimination of this position when the valve port is not in use. It is recommended that a six-inch length of hose, plugged at the outer end, be attached to the bomb bay valve ports.

f. *Turret Operation.*

(1) *General.*

(a) Displacement of the guns in elevation is controlled by lifting or depressing the hand control grips (figure 72-3). The direction of such displacement (increasing or decreasing) corresponds to the direction of the handgrip motion about the horizontal axis.

(b) Rotation of the turret is obtained by turning the handgrips about the vertical axis. The range knob (figure 72-2) is mounted conveniently between the grips, so that the gunner rests both thumbs on this knob while holding the grips in the palms of his hands. This knob sets the range in the computing sight.

(c) The hydraulic power unit furnishes the mechanical power for driving the turret in azimuth and the guns in elevation.

(d) A gun firing switch is mounted to the rear and at the upper end of each handgrip. The two firing switches are connected in parallel so that either switch can be used to fire the guns. A dead-man switch is provided on each grip; these are connected in parallel so that the gunner can operate the turret when either hand rests on a grip. The dead-man switch is provided so that in event of mishap to the operator, causing his hands to fall off the grips, the power circuits of the turret will be opened and all turret motion and firing of guns will be stopped.

(2) *Preflight Operation.*

(a) Allow hydraulic units and sight to warm up at least five minutes before take-off.

(b) Engage power clutches.

(c) See that hand cranks are disengaged (do not disengage until after power clutches have been engaged).

(d) Feed ammunition just up to the guns.

(e) Move main gun switch (figure 74-1) to "ON" position.

(f) Place sight switch (figure 67-3) in "ON" position.

(g) Close dead-man switches (figure 72-1) on handgrips.

(h) Check response of azimuth and elevation mechanisms by manipulating the handgrips. (See figure 72-3.)

(i) Turn range knob (figure 72-2) and observe that reticles move in response.

(j) Adjust reticle light (figure 67-2) to desired brilliance (approximately).

(3) *Flight Operation.*

(a) Charge guns by pulling each handle (figure 67-4) twice.

(b) Turn on gun selector switches. (See figures 74-2 and 74-3.)

(c) When target is sighted, set in target dimension on sight.

(d) Turn hand controls (figure 72-3) so that reticles frame the target.

(e) Adjust range knob (figure 72-2) until reticles frame the target.

(f) Press either firing switch.

(g) After ammunition has been used, charge guns at least twice to clear out live shells.

(h) When the turret is not being used, turn it so that the guns point aft and are parallel to the center line of the airplane.

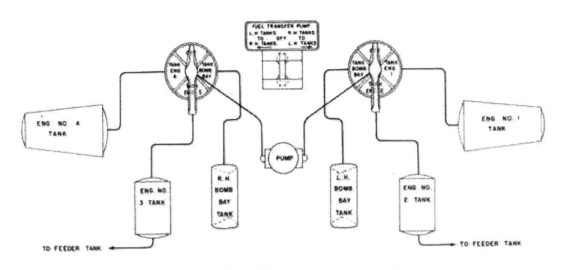

Figure 71 - Fuel Transfer Operation Diagram

- 114 -

RESTRICTED

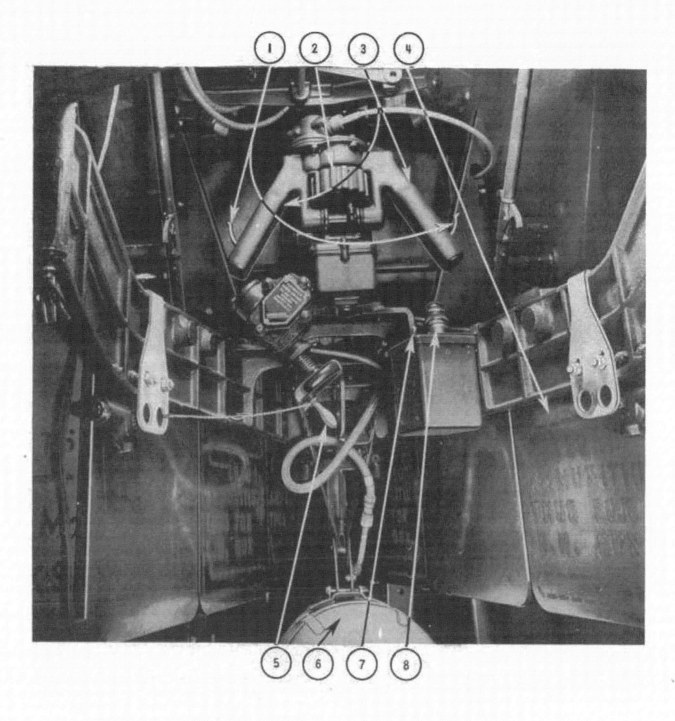

1 Deadman Switch
2 Range Knob
3 Hand Grip
4 Ammunition Box
5 Azimuth Handcrank
6 Oxygen Bottle
7 Trouble Light Switch
8 Trouble Light

Figure 72 - Upper Turret Controls

1 Range Knob
2 Trouble Light Switch
3 Trouble Light
4 Hand Grip
5 Deadman Switch
6 Oxygen Flow, Control Knob
7 Oxygen Mask Connection Fitting
8 Elevation Handcrank

Figure 73 - Oxygen Regulator-Upper Turret

(i) In event of power failure, the turret may be controlled by the azimuth hand crank (figure 72-5) and the elevation hand crank (figure 73-8) as set forth in the following paragraph. It is not possible to track a target when the hand cranks are used, since the movement of the guns and turret would be too irregular for accurate sighting even if the gunner could manipulate all the controls at one time. The hand cranks may be used, however, for positioning the turret and guns so that they point to the approximate position of the target.

(j) When it is necessary to use the hand cranks, the following procedure should be followed exactly:

1. Engage azimuth (figure 72-5) and elevation hand cranks (figure 73-8).

2. Disengage power clutches.

3. Move turret and guns into desired position.

4. When finished, re-engage power clutches.

5. Be sure to disengage hand cranks before operating power motor again.

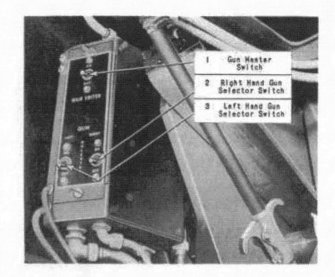

Figure 74 - Upper Turret Gun Selector Switches

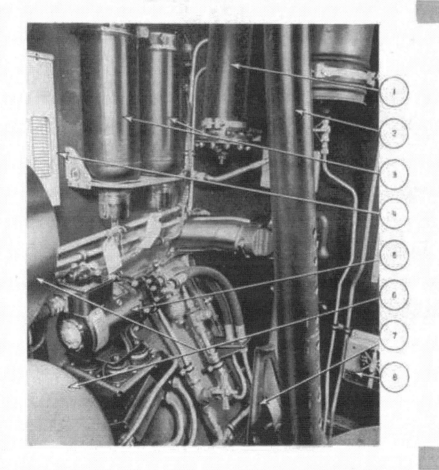

1 Auxiliary Hydraulic Accumulator
2 Turret Support Structure
3 Thermos Bottles
4 Paper Cup Dispenser
5 Electric Hydraulic Pump
6 Oxygen Bottles
7 Top Gunner's Foot Rest
8 Fuel Tank Transfer Selector Valve

Figure 75 - Upper Turret Compartment - Right Side - Looking Aft

1 Copilot's Armor Plate Support
2 Fire Extinguisher
3 Inverter
4 Paper Cup Dispenser
5 Wiring Diagram Holder
6 Oxygen Bottles

Figure 76 - Upper Turret Compartment - Right Side - Looking Forward

SECTION VIII

OPERATING INSTRUCTIONS - BALL TURRET COMPARTMENT

1. General Description.

a. In the bottom of the fuselage between the side windows and the radio compartment is installed a Sperry ball-type power turret equipped with twin .50 caliber machine guns. Entrance into the turret is accomplished from within the airplane after take-off. An auxiliary oxygen bottle is fastened to the upper support of the turret and is not part of the airplane's master oxygen system. A first aid kit is clipped to the aft side of the bulkhead between the ball turret compartment and the radio compartment to the left of the door.

b. A hydraulic power unit provides power for driving the turret in azimuth and elevation. It provides the gunner with the power for obtaining smooth and easily controlled motion of the turret when tracking the target.

c. The hand control and limit unit controls the outputs of the azimuth and elevation hydraulic systems. A pair of handgrips (figure 86-2) controls the motion of the turret in azimuth and elevation. Each handgrip has a firing switch on the top end. (See figure 86-3).

d. The switch box (figure 85-7) controls distribution of the electric power to the various units in the turret. The terminal block in the top left end of the box has convenient posts for connecting the leads of the gunner's head set and microphone.

e. A press-to-talk switch (figure 77-1) for intercommunication is located just in front of the gunner's right footrest.

f. A rheostat control is provided for use with the gunner's heated suit. It is located on the underneath side of the seat and is adjusted to obtain the desired temperature in the suit.

2. Operational Equipment.

a. Lighting. - A dome light is located in the ceiling of the compartment just aft of the turret support. The switch (figure 78) is located to the right of the door between this compartment and the radio compartment.

Figure 78 -
Dome Light Switch
Ball Turret Compartment
Looking Forward

Figure 79 -
Refilling Ball Turret
Oxygen Bottle

b. Oxygen Controls. - An oxygen regulator (see inset) is provided on the inside of the ball turret on the right-hand side. Oxygen is supplied from the auxiliary bottle above the turret. When the supply of this auxiliary bottle is exhausted, it can be renewed from the airplane's master system by means of a filler valve connection as illustrated in figure 79. Operation of the regulator is conventional.

c. Armor Protection. - Refer to figure 80 for angles of armor protection.

d. Operation of Turret.

(1) Instructions for Entering Turret.

CAUTION: Do not attempt to rotate the turret in elevation while the airplane is on the ground.

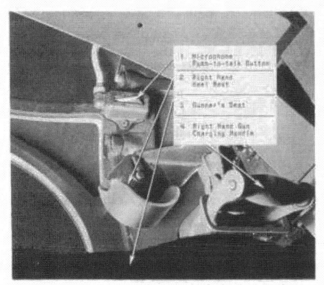

Figure 77 - Ball Turret Gunner's Microphone Button

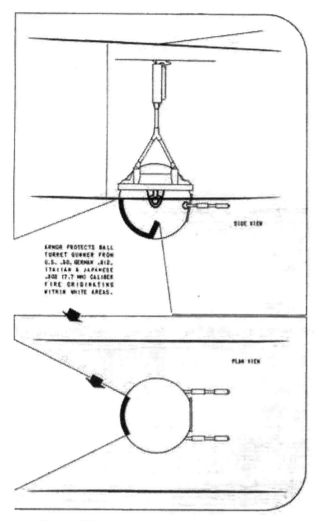

Figure 80 - Angles of Armor Protection Diagram - Ball Turret

No crew member shall be in the turret during landing or take-off and the guns of the turret shall be in the horizontal position pointing aft.

(a) Remove ammunition box cover and load. Push ammunition down to the guns.

(b) Remove elevation hand crank (figure 81-1) from its clip and attach it to shaft. Be sure that the hand brake (figure 81-2) is locked.

(c) Move elevation hand clutch (figure 81-4) to "IN" position. It may be necessary to loosen hand brake and rock hand crank back and forth before hand clutch can be moved to "IN" position.

(d) Move elevation power clutch to "OUT" position using clutch handle, then replace handle in its clip.

(e) Loosen elevation brake (figure 81-2) slowly while holding elevation hand crank firmly.

(f) Turn elevation hand crank (figure 81-1) in down direction as illustrated in figure 82 until turret revolves to low limit of elevation (-90 degrees).

(g) While holding elevation hand crank, open turret door, reach inside and move elevation power clutch to "IN" position.

(h) Move elevation hand clutch to "OUT" position, remove hand crank and replace it in its clip.

(i) Enter turret. Close door securely. Be sure door handles are pushed all the way up and that the turret door is locked before turning main power and sight switches "ON."

(2) Preflight Operation.

(a) Turn power switch "ON."

(b) Turn sight switch (figure 83-3) "ON."

(c) Check response of azimuth and elevation mechanisms by manipulating the hand controls.

WARNING: Be sure that the guns are not driven down into the ground.

(d) Adjust reticle light (figure 83-2) on sight to desired brilliance (approximately).

(e) Work range foot pedal (figure 85-4) and observe if reticles move in response.

(f) Lift each gun cover plate and pull ammunition down, feeding first shell by hand into magazine of gun, then close gun cover plates.

(3) Flight Operation.

(a) Load ammunition boxes as illustrated in figure 83. Enter turret.

(b) Turn on power switch.

(c) Turn on sight switch. (See figure 83-3.)

(d) Charge guns by pulling charging handles (figure 85-3) twice.

(e) Turn on fire selector switches. (See figure 86-9.)

(f) By means of hand controls (figure 86-2), track the target.

(g) Operate range foot pedal (figure 85-4) until reticles frame the target.

(h) Close either firing key. (See figure 86-3.)

(i) When ammunition is used up, charge guns at least twice to be sure that no live shells are left in the guns.

Figure 81 - External Manual Controls - Ball Turret

1 Elevation Handcrank
2 Elevation Handbrake
3 Lug Wrench
4 Elevation Handclutch

Figure 82 - Operation of External Elevation Handcrank - Ball Turret

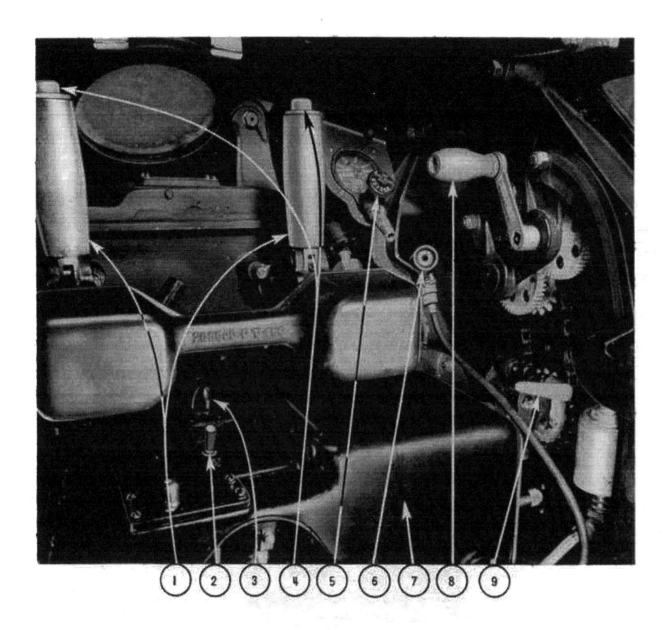

1 Hand Grips
2 Sight Light Rheostat Control
3 Gun Sight Switch
4 Gun Firing Switches
5 Oxygen Regulator
6 Flexible Light
7 Gun Sight
8 Azimuth Hand Crank
9 Azimuth Power Clutch

Figure 83 - Ball Turret Azimuth Handcrank

RESTRICTED

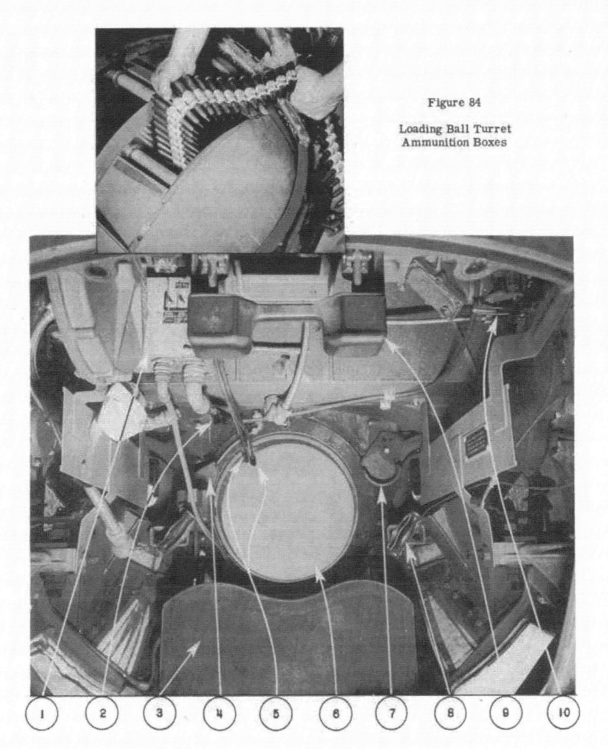

Figure 84

Loading Ball Turret Ammunition Boxes

1 Electrical Switch Box
2 Spotlight Switch
3 Gunner's Seat
4 Range Foot Pedal
5 Headset and Microphone Leads
6 Turret Front Window
7 Foot Rest
8 Charging Handle
9 Turret Hand Control & Limit Unit
10 Elevator Power Clutch

Figure 85 - Ball Turret - Top View

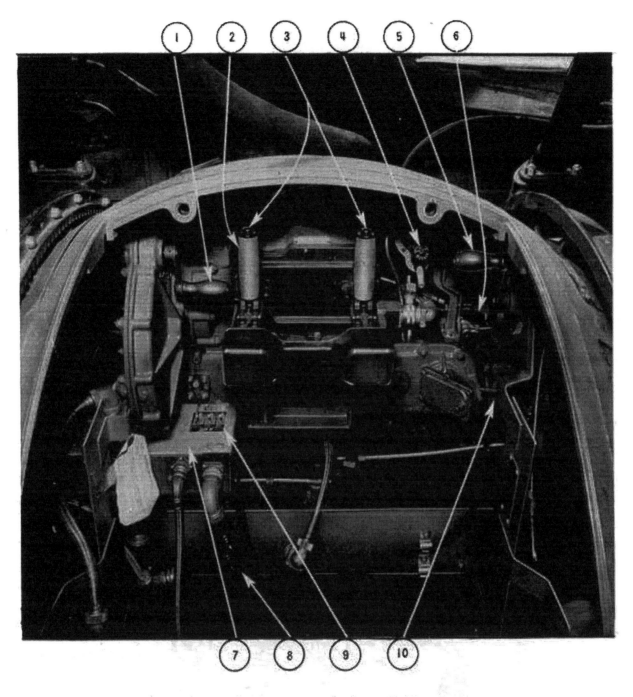

1 Elevation Handcrank
2 Hand Control Grip
3 Firing Switches
4 Oxygen Regulator
5 Azimuth Handcrank
6 Spot Light
7 Electrical Switch Box
8 Spot Light Control Switch
9 Gun Selector Switches
10 Elevation Power Clutch

Figure 86 - Ball Turret Controls

(j) Turn fire selector switches (figure 86-9) "OFF."

NOTE: Hand cranks are provided on the interior of the ball turret for manual operation of the turret in case of power failure. The azimuth hand crank connection is on the right side (see inset) and the elevation hand crank connection (figure 87) is on the left. The two hand cranks are stowed in clips just above the door opening.

(4) <u>Instructions for Leaving Turret</u>.

(a) Drive turret to low limit of elevation (-90 degrees).

(b) Turn main power and sight switches (figure 86-9) "OFF."

(c) Open door and leave turret.

(d) Attach elevation hand crank. (See figure 81-1.)

(e) Move elevation hand clutch (figure 81-4) to "IN" position.

(f) While holding elevation hand crank firmly, reach inside turret and move elevation power clutch to "OUT" position.

(g) Close and latch door of turret.

(h) Turn the elevation hand crank and elevate the turret to upper limit, and then lock elevation hand brake. (See figure 81-2.)

(i) Remove clutch handle from clip and attach to <u>azimuth</u> power clutch on top of turret. Move azimuth power clutch to "OUT" position and push turret by hand until guns point to rear of airplane.

(j) Move azimuth power clutch to "IN" position, and then remove handle and replace it in its clip.

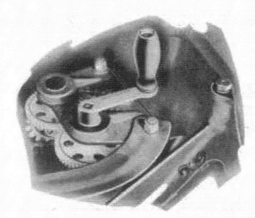

Figure 87 - Ball Turret Elevation Handcrank

SECTION IX

OPERATING INSTRUCTIONS - SIDE GUNNER'S COMPARTMENT

1. **General Description.**

   The side gunner's compartment extends from the bulkhead just aft of the main entrance door forward to the lower ball turret. It includes storage space for the nacelle platforms, one flexible .50 caliber machine gun at each side window, a chemical toilet just aft of the main entrance door, a fire extinguisher and oxygen and interphone controls.

2. **Operational Equipment.**

   a. *Initial Entrance.* - Through the main entrance door. (See figure 88-2.)

   b. *Lighting.* - The dome light switch (figure 88-3) is located just aft of the main entrance door.

   c. *Interphone Controls.* - The interphone jackbox (figure 89-2) is located on the left side of the compartment, on the forward side of the left-hand gun support. Operation is conventional.

   d. *Suit Heater Control.* - A rheostat control (figure 89-4) is provided for use with the gunner's heated suit. It is adjusted to obtain the desired temperature in the suit.

   e. *Oxygen Controls.* - The two regulators (figure 90) are located side by side in the ceiling between the two guns.

   f. *Emergency Equipment.*

   (1) *Fire Extinguisher.* - A carbon tetrachloride fire extinguisher (figure 88-5) is attached to the forward side of the bulkhead between the tail wheel compartment and the waist gunner's compartment, behind the chemical toilet.

   (2) *Emergency Releases.* - Each side window has an emergency release bar (figures 92-4 and 93-2) on the forward side of each window. To open the window, jerk the bar forward. There are no catches to be released. The main entrance door also has an emergency release handle. (See figure 88-1.)

   g. *Armor Protection.* - For angles of armor protection, see figure 91.

   h. *Flight Operation.* - To prepare the machine guns for action, remove the straps (figures 92-2 and 93-3) and swing the guns into position. Loading and firing the guns is conventional.

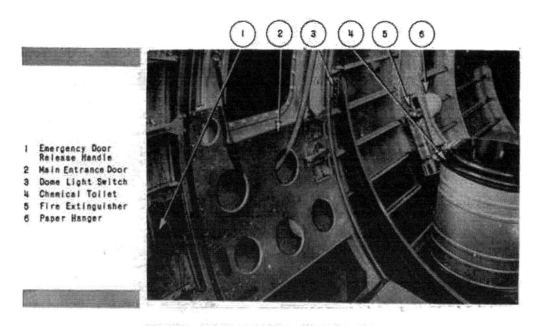

1 Emergency Door Release Handle
2 Main Entrance Door
3 Dome Light Switch
4 Chemical Toilet
5 Fire Extinguisher
6 Paper Hanger

Figure 88 - Main Entrance Door - Side Gun Compartment

1 Phone Call Lamp
2 Interphone Jackbox
3 Ammunition Boxes
4 Suit Heater Outlet
5 Oxygen Filler Valve

Figure 89 - Side Gunner's Controls

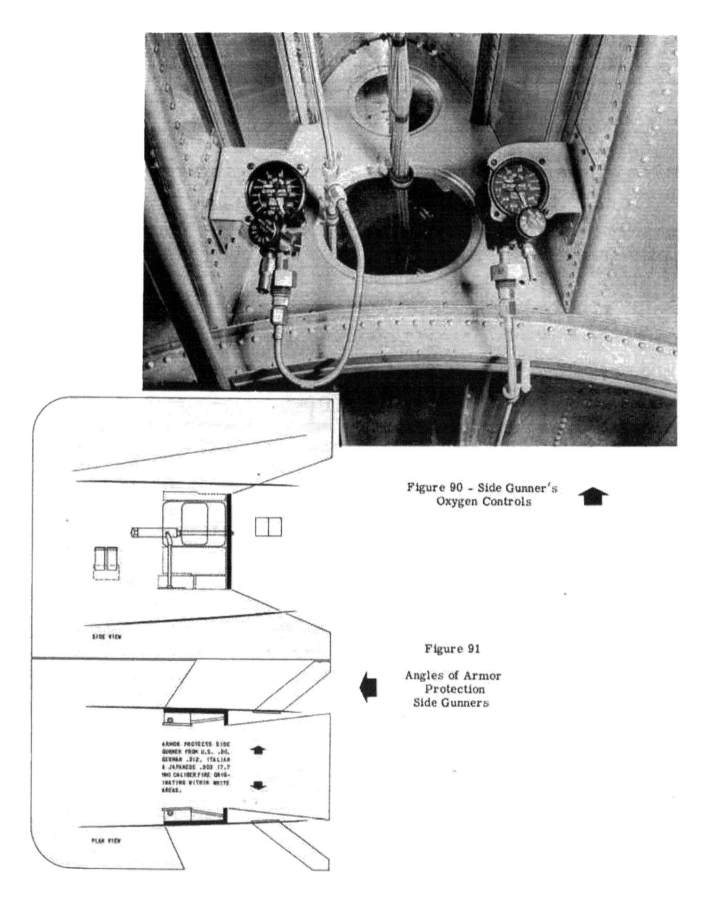

Figure 90 - Side Gunner's Oxygen Controls

Figure 91

Angles of Armor Protection Side Gunners

1. Ammunition Box
2. Stowage Strap
3. Armor Plate
4. Bar For Opening Window
5. .50 Caliber Machine Gun
6. Phone Call Lamp
7. Interphone Jackbox
8. Ammunition Boxes

Figure 92 - Side Gun Compartment-Left Hand Side-Looking Aft

1. .50 Caliber Machine Gun (Stowed)
2. Bar for Opening Window
3. Stowage Strap
4. Ammunition Boxes
5. Armor Plate
6. Fire Extinguisher
7. Chemical Toilet

Figure 93 - Side Gun Compartment - Right Hand Side - Looking Aft

SECTION X

OPERATING INSTRUCTIONS - BOMB BAY COMPARTMENT

1. **General Description.**

   The bomb bay is situated between the radio compartment and the upper turret compartment. It is equipped with either bomb racks or releasable self-sealing fuel tanks, two dome lights, step light, an oxygen regulator, relief tube located behind the left-hand dome light, connections for manual operation of the landing wheels separately and bomb bay doors, and an emergency bomb release handle. For detailed operating instructions on the use of the hand fuel pump, see figure 98A.

2. **Operational Equipment.**

   a. **Lighting.**

   (1) The step light (figure 94-4) is in the step at the forward end of the catwalk. The switch (figure 50-2) is on the forward wall of the radio compartment to the right of the door.

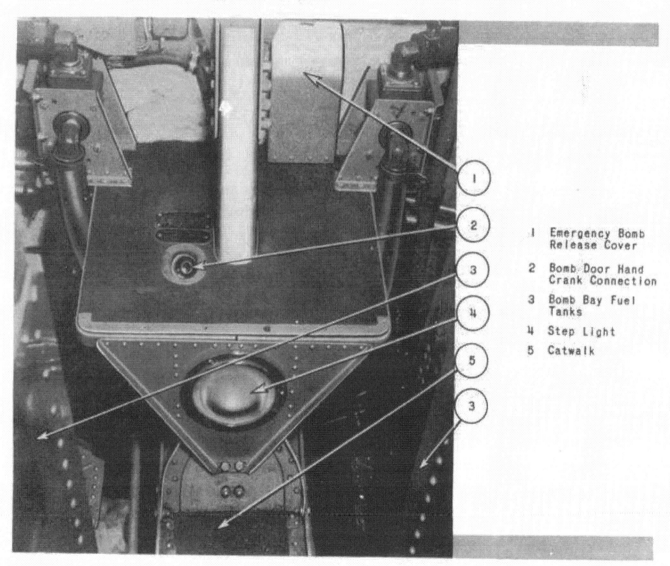

1 Emergency Bomb Release Cover
2 Bomb Door Hand Crank Connection
3 Bomb Bay Fuel Tanks
4 Step Light
5 Catwalk

Figure 94 - Step - Bomb Bay - Looking Forward

RESTRICTED T.O. NO. 01-20EF-1

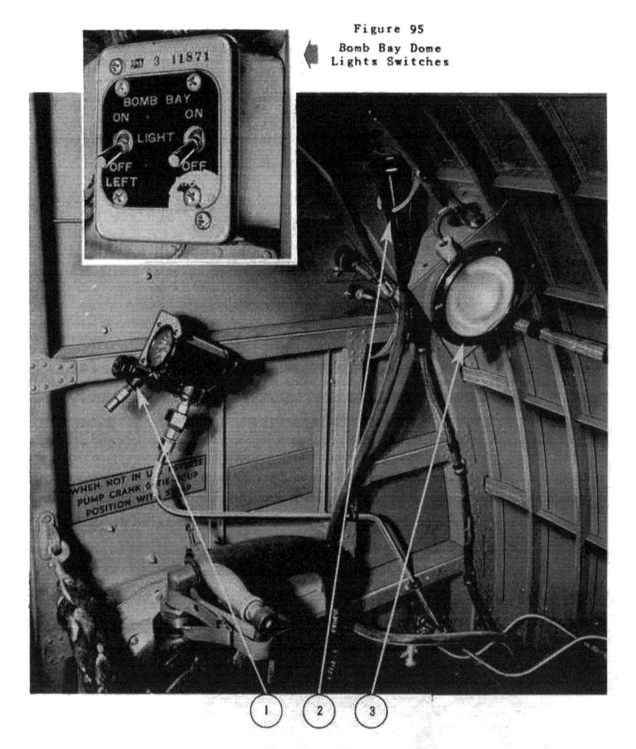

Figure 95
Bomb Bay Dome Lights Switches

1 Oxygen Regulator
2 Relief Tube
3 Left Hand Dome Light

Figure 96 - Oxygen Regulator - Bomb Bay

RESTRICTED

(2) Two dome lights (figure 96-3) are located in the bomb bay, one on either side at the aft end of the bay. The switches (figure 95) are located on the forward side of the aft bulkhead to the right of the door.

b. Oxygen Regulator. - The oxygen regulator (figure 96-1) is located on the aft wall of the bomb bay on the left side of the compartment.

c. Emergency Equipment.

(1) Emergency Operation of Landing Wheels. - A hand crank connection for manual operation of each wheel is located on each side of the door in the forward wall of the bomb bay. Two hand cranks are clipped to the aft wall of the radio compartment. Insert the crank into the connection and rotate clockwise to raise the wheel and counterclockwise to lower it. (See figure 115.)

(2) Emergency Operation of Bomb Bay Doors. - Use the same crank for manual operation of the bomb bay doors that was used for the manual operation of the landing wheels. Insert the crank into the connec-

Figure 98 - Right Hand Bomb Rack Selector Switch

tion (figure 94-2) in the step at the forward end of the catwalk in the bomb bay. Rotate clockwise to close the door and counterclockwise to open them.

(3) Emergency Bomb Release. - The emergency bomb release handle is located on the step at the forward end of the bomb bay and is protected from inadvertent handling by a hinged guard (figure 94-1) which may be moved out of position when desired. Pulling the handle will result in release of all bombs salvo the instant the doors reach the full open position. The bomb bay fuel tanks may also be dropped by the emergency release handle. For instructions on retraction of the bomb doors after emergency release, see figure 119.

d. Bomb Rack Selector Switches. - Two switches, one on each side of the bomb bay, are used in conjunction with the two bomb rack selector switches mounted on the bombardier's control panel. (See figure 63-17.) When either switch is in the "OFF" position, electrical release of bombs, or fuel tanks, from that bay is impossible.

Figure 97 - Left-Hand Bomb Rack Selector Switch

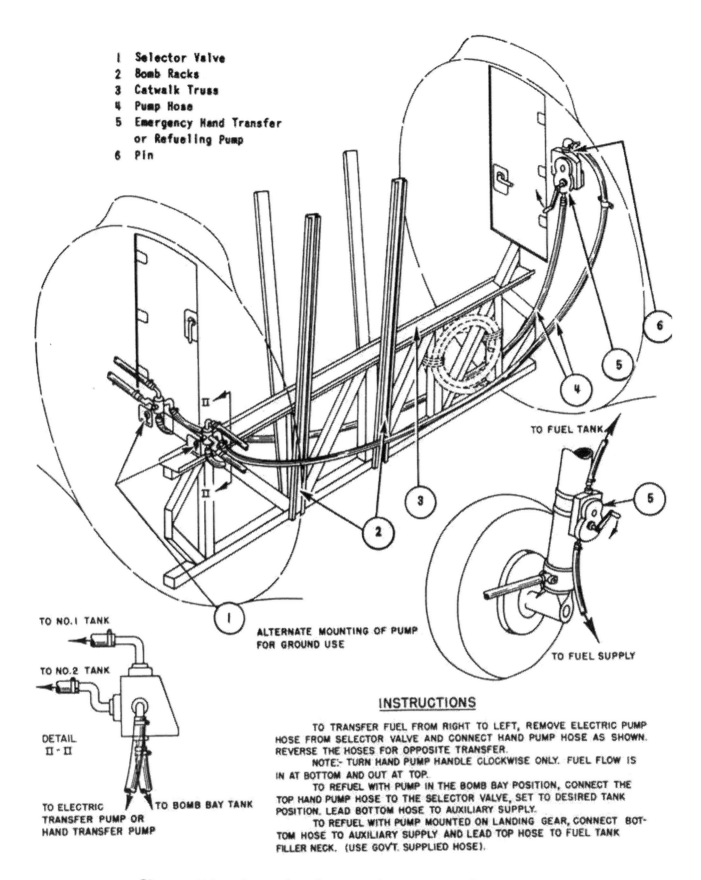

Figure 98A – Operating Instructions – Hand Fuel Pump

SECTION XI

OPERATING INSTRUCTIONS - TAIL GUNNER'S COMPARTMENT

1. **General Description.**

The tail gunner's compartment is located in the tail of the airplane. It is equipped with a side door, two oxygen regulators, interphone controls, a dome light and switch, twin .50 caliber machine guns, bicycle-type seat, knee pads, phone call lamp and gun camera, mounted between the gun barrels, with safety switch.

2. **Operational Equipment.**

   a. *Initial Entrance.* - There are two ways of entering the tail gunner's compartment: one from the tail wheel compartment through a small door in the bulkhead, and one from the outside through a side door. (See figure 100) The latter is used for emergency exit, and is equipped with an emergency release handle. (See figure 100)

   b. *Lighting.* - A dome light and switch are located just above the gun handles behind the armor plate.

   c. *Seat Adjustment.* - For complete instructions for adjusting the tail gunner's seat, see figure 101.

   d. *Interphone Jackbox.* - The jackbox (figure 102-1) is located on the right side of the compartment looking aft above the aft end of the ammunition box. Operating instructions are outlined in section I, paragraph 4.j.

   e. *Oxygen Controls.* - Two oxygen regulators (figures 103 and 104-3) are provided, one on each side wall. Operation is conventional.

   f. *Armor Protection.* - The window through which the gunner sights is made of bullet resistant glass. For angles of armor protection, see figure 105.

   g. *Suit Heater Control.* - A rheostat control (figure 106) is provided for use with the gunner's heated suit. It is adjusted to obtain the desired temperature in the unit.

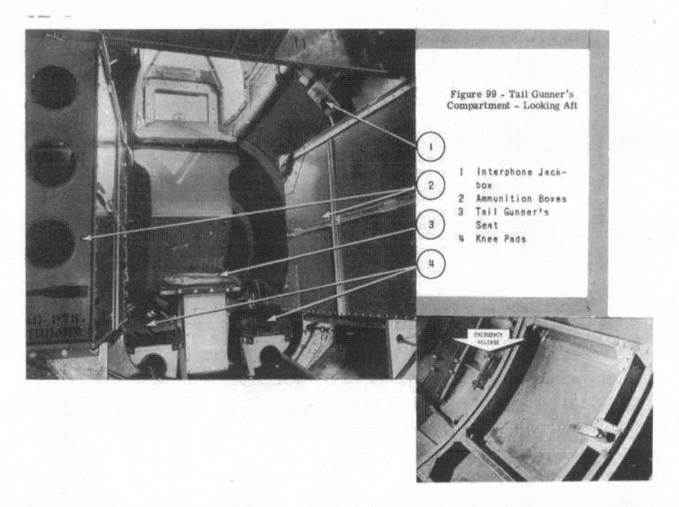

Figure 99 - Tail Gunner's Compartment - Looking Aft

1. Interphone Jackbox
2. Ammunition Boxes
3. Tail Gunner's Seat
4. Knee Pads

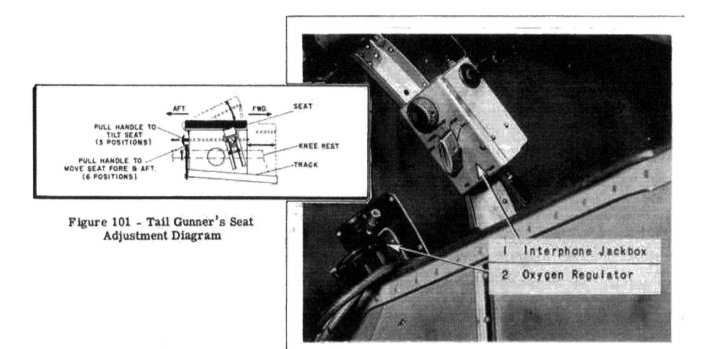

Figure 101 - Tail Gunner's Seat Adjustment Diagram

Figure 102 - Left Side Wall Controls - Tail Gunner's Compartment

Figure 103 - Tail Gunner's Phone Call Lamp

Figure 104 - Right Side Wall Controls - Tail Gunner's Compartment

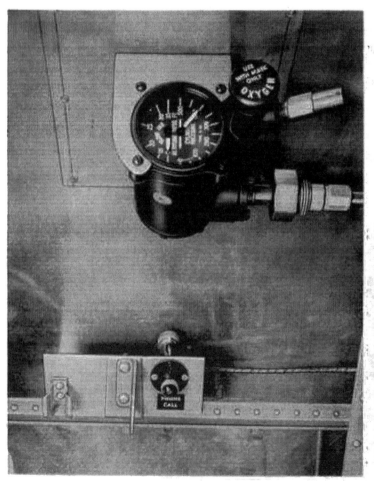

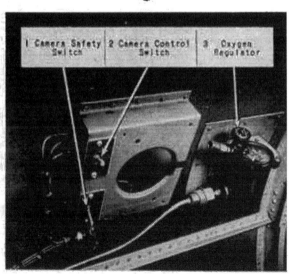

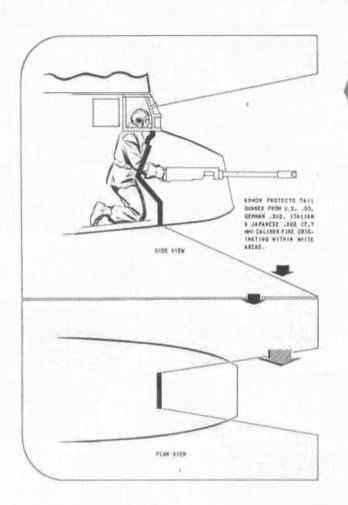

Figure 105 - Angles of Armor Protection Diagram Tail Gunner's Compartment

Figure 106 - Tail Gunner's Suit Heater Outlet Right-Hand Side

Figure 107 - Tail Wheel Compartment Looking Aft

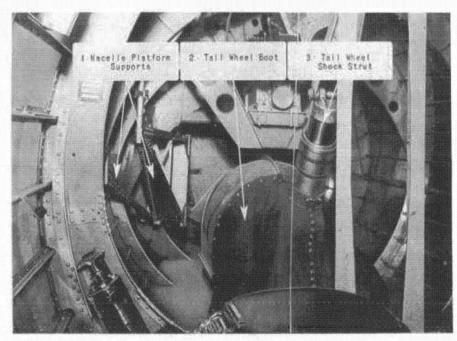

SECTION XII

OPERATING INSTRUCTIONS - TAIL WHEEL COMPARTMENT

1. General Description.

The tail wheel compartment is located between the side gunner's compartment and the tail gunner's compartment. It is equipped with a boot (figure 107-2) for the tail wheel, nacelle platform supports (figure 107-1), a connection for manual operation of the tail wheel, and an alarm bell. (See figure 121.)

2. Operational Equipment.

a. **Lighting.** - There are no lights located in this compartment. Light is derived from the dome light in the side gunner's compartment, which is turned on by the switch (figure 88-3) located just aft of the main entrance door.

b. **Emergency Operation of Tail Wheel.** - The hand crank for manual operation of the tail wheel is located in the radio compartment clipped to the right-hand aft wall of the compartment, above the transmitter tuning units. The crank is placed into the connection (figure 116) which is located directly behind the electric retraction motor. Rotate as desired.

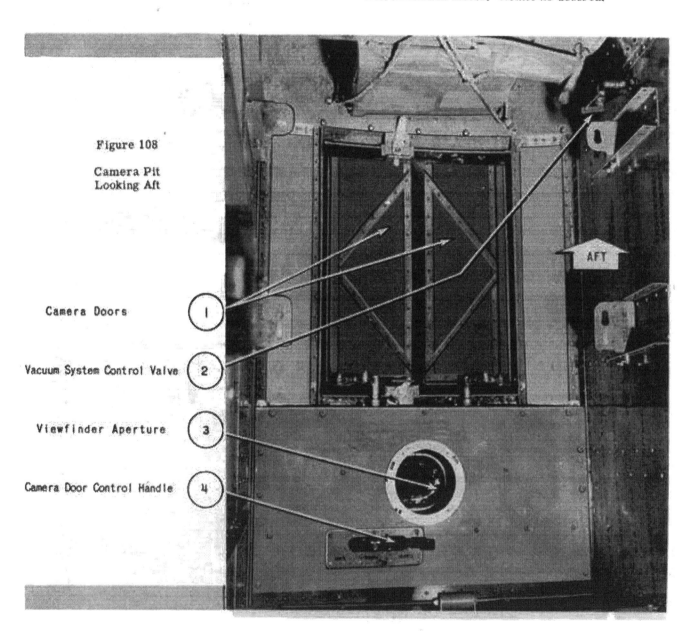

Figure 108

Camera Pit Looking Aft

1. Camera Doors
2. Vacuum System Control Valve
3. Viewfinder Aperture
4. Camera Door Control Handle

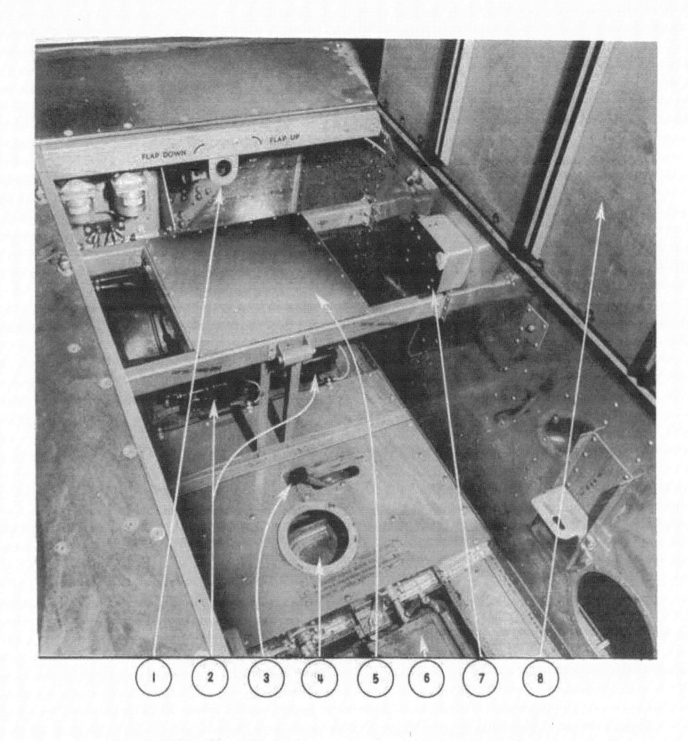

1. Wing Flap Hand Crank Connection
2. Propeller Anti-Icer Pumps
3. Camera Door Control Handle
4. Viewfinder Aperture
5. Camera Operator's Seat
6. Camera Door
7. Intervalometer Power Receptacle
8. Camera Pit Door

Figure 109 - Camera Pit-Looking Forward

RESTRICTED

SECTION XIII

OPERATING INSTRUCTIONS - CAMERA PIT

1. **General Description.**

   Camera equipment is installed in the pit under the door of the radio compartment. A door (figure 109-8) is provided in the floor for the use of the equipment by the camera operator. Provision is made for three alternate installations as follows:

   Type T-3A Installation:

Camera	type T-3A
Camera Mount	A-5A
View Finder	A-2
Filter	A-3
Shutter Induction Coil	

   Type K-3B Installation:

Camera	type K-3B
Camera Mount	A-8
View Finder	A-2
Intervalometer	
Magazine	A-1A
Filter	A-2A

   Type K-7C Installation:

Camera	type K-7C
Camera Mount	A-8
View Finder	A-2
Filter	A-4

2. **Operational Equipment.**

   a. The type A-2 view finder may be installed forward of the camera. The bracket assembly used to support the intervalometer is stowed on the right side of the camera pit. The intervalometer is stowed on the right side. A direct current power receptacle for the intervalometer is installed on the right-hand side of the pit and a connection to the vacuum system is provided on the left-hand side.

   b. The double camera doors (figure 108-1) and the view finder door (figure 108-3) are hinged in the bottom of the fuselage and are operated by a lever (figure 109-3) located on the floor at the operator's seat (figure 109-5).

SECTION XIV

COLD WEATHER OPERATION

3. **Winterization.**

   When the outside temperature is expected to fall below 0° C (32° F), the airplane should not be operated until "winterization" service has been completed. The pilot should definitely ascertain that this service has been completed before attempting to operate the airplane or any component part thereof. The following major items are listed with recommended disposition for a change over from warm weather operating service.

Self-thawing engine oil cooling system	O.K.
Oil radiator shutters	O.K.
Oil dilution system	O.K.
Carburetor heat 32.2° C (90° F) or alcohol de-icing system.	X
Propeller anti-icer system	X
Battery cart external plug	O.K.
Cockpit heating system (including windshield defrosters)	X
Snow and ice tread tires	S.C.
Wing and engine covers, engine heaters	X
Non-channeling grease in control systems	O.K.
Lubricate propeller with special grease	S.C.
Lag front row radial engine push rod housing	X
Long reach hot running spark plugs	S.C.
Immersion heaters in oil tank	X
Fuel tank drain	O.K.
Engine cowl flaps	X
Light oil in oil pressure gage line	X&S.C.
High capacity ignition system	X
Carburetor air thermometers	X
Change hydraulic oil	S.C.
Use winter grade engine oil (special)	S.C.
Wing de-icing equipment	O.K.
Full closing intercooler shutters	O.K.
Supercharger regulator and its oil system	X
Windshield wipers	X
Windshield de-icers	O.K.
Oil tank sump drain, 770-1 drain cock	X
Provisions for priming with high volatile fluids	X
Lag oil lines and tank where non-self-sealing	X

   Legend:

O.K.	Satisfactory as is.
X	This item requires action.
S.C.	Service Change.

2. **Engine Oil Dilution System.**

   a. **Controls and Indicators.**

   (1) The oil dilution system provides a method of diluting or thinning the engine oil with gasoline at the end of each engine run in order to facilitate starting the engine in cold weather.

   (2) The system consists of four electric solenoid-operated oil dilution valves, each located on the front of its respective engine fire wall. The necessary piping, wiring, and four toggle switches are on the copilot's control panel.

   (3) The engine oil should be diluted prior to stopping the engines, when there is a possibility of the engine oil temperature dropping below approximately 5° C (41° F) during the period the engine is to be inoperative.

   b. **To Dilute Oil.**

   (1) Maintain a speed of 800 rpm for each engine. If an engine speed in excess of 800 rpm is maintained, the oil temperature will exceed the maximum temperature limit set for the diluting period. Fuel vapor blown from the breather outlets to the exhaust stacks by the propeller blast also creates a fire hazard.

   NOTE: It is impossible to dilute engine oil unless the engine is running.

   (2) Maintain the oil temperature of each engine below 50° C (122° F) during the dilution procedure. The ideal temperature is 40° C (104° F). If the oil temperature exceeds 50° C (122° F), the gasoline will evaporate as rapidly as it is introduced into the oil and will leave the oil with its original viscosity. This vaporizing fuel exhausting from the breather outlets creates a dangerous fire hazard. If the temperature exceeds 50° C (122° F) when the airplane is landed, the engines must be stopped and the oil allowed to cool to approximately 35° C (95° F) before the engines are started again to accomplish oil dilution.

   (3) Hold the oil dilution switch in the "ON" position for 4 minutes plus the time required for the propellers to stop rotating. The engines must be stopped at the end of the dilution period by moving the mixture control (figure 40-7) to the "IDLE CUT-OFF" position. The fuel pressure should show a drop from normal pressure to approximately 4 or 5 pounds per square inch during oil dilution. If a sharp decrease in fuel pressure is not noted, check the oil dilution electrical circuits, the oil dilution valves, and the pressure gages for the source of the trouble.

   (4) During oil dilution period, the supercharger controls should be operated continuously to expedite the flow of diluted oil to the supercharger regulators. With warm oil in the engine the minimum time for operating the regulator control from the low boost to the high boost positions should be 5 seconds. If the oil is somewhat cooler than normal engine temperatures, this time should be extended to 15 seconds.

   (5) The electric booster pumps need not be running during the dilution period.

   (6) When the engines are started subsequent to engine oil dilution, normal starting procedure, as outlined in section II, paragraph 3., should be followed.

   (7) The copilot's instrument panel carries dual oil pressure (figure 38-2) and temperature gages (figure 38-9). Pressure is measured at the pressure side of the oil pump. The oil temperature is that of the oil entering the engine and is taken at the "Y" valve when the oil leaves the tank.

3. **Propeller Oil Dilution.**

   When operating in cold climates with oil dilution equipment, the propeller control will be moved slowly from extreme increase to extreme decrease rpm several times during the period of dilution. This will permit the filling of the dome of the propeller with diluted oil and prevent sluggish response of the propeller when starting the engine the next time.

4. **Portable Ground Heaters.**

   a. When operating under freezing conditions, and if available, use type D-1 portable heaters. (See figure 110.) The weather conditions may require preheating of engines and cabins prior to first flight.

Figure 110 - Type D-1 Portable Ground Heater

   b. It requires approximately 15 minutes to heat up the engines at -17.8° C (0° F) and approximately 30 minutes at -34.4° C (-30° F). Each heater is normally equipped with three flexible warm air ducts.

   c. One heater may be used in light freezing weather to heat two engines and the cabin at a time. Extreme cold weather conditions might require that the entire

Figure 111 - Handling of Type D-1 Portable Ground Heater

output of one heater be directed into each engine and one into the cabin. The heater weighs approximately 210 pounds, is equipped with two rubber wheels and is easily handled by one man. (See figure 111.) Detailed operating instructions are contained in figure 112.

CAUTION: Whatever method is used for preheating the engine, extreme care must be taken to prevent accidental ignition of the gas fumes from the engine breathers, caused by vaporization of the gasoline in the oil.

5. Cold Weather Starting of Engines.

a. When the engines are to be started for warm-up, or to be repeatedly started and stopped for ground test purposes or "alert," engines will be primed and the oil dilution system operated in accordance with instructions given in paragraph 1.a., this section.

b. During cold weather operation, drain the oil pressure gage line and refill with instrument oil.

WARNING: In warming a cold engine in extremely cold weather, start with cowl flaps closed. Do NOT gun engine to more than 900 rpm until oil has reached a temperature of 40° C (104° F).

6. Batteries.

Energizers or battery carts are generally used for cold weather starting, as this is more practicable than heating the batteries. Batteries should be maintained at not less than -12.2° C (10° F). Lower voltage at extremely low temperatures causes malfunctioning of all electrical equipment.

NOTE: To safeguard batteries, remove them from the airplane and store them in a heated place when the airplane is to be idle overnight.

Figure 112 - Detailed Instructions - Portable Ground Heater

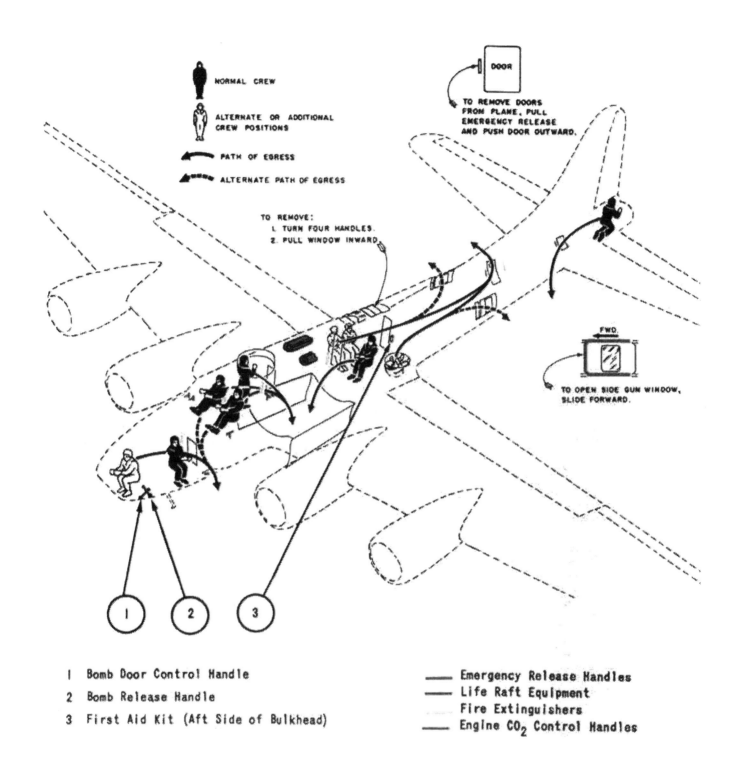

Figure 113 - Emergency Equipment and Exits Diagram

7. Protective Covers.

Airplane and engine protective covers, consisting of tarpaulins with ropes for securing the airplane, will be used on the wings, tail surfaces, and fuselage to provide protection against frost accumulation. Engines and compartment enclosures should be covered also.

8. Frost or Ice Removal.

When it is necessary to remove frost or ice from areas of the airplane, melt a small area of the ice-covered surface at a time using hot water, then flush this area with denatured alcohol before the hot water freezes. Pay particular attention to the hinges and controls. Alcohol should be used for cleaning frost off windows and windshields.

9. Mooring.

If, due to extreme cold weather, mooring stakes cannot be driven into the ground, use a pick or other sharp instrument, and dig a hole (approximately 8 inches deep and 8 inches square). Into this hole place deeply notched stakes crosswise, and then tie the mooring rope to the stakes. Fill the hole with water which will freeze the stakes and rope fast. If stakes are unavailable, dig the hole, and coil the rope in the bottom of it, then fill the hole with water. Moor the airplane with the nose into the wind.

10. Communications Equipment.

The following equipment is adversely affected by extreme cold weather:

a. Dynamotor. - The increased viscosity of bearing lubricants may prevent the dynamotor from starting resulting in blown fuses. If this occurs the grease should be removed and oil substituted as a lubricant.

b. Operating Controls, Hand Switches, Etc. - Stiffness of operation may occur. Oil should be removed in order to prevent drag and binding.

c. Storage Batteries. - Cracking occurs around the edge of the case. Batteries should be kept charged above 1.290 specific gravity to prevent cracking.

d. Microphones. - The hand microphone is unsatisfactory for use in cold weather. Moisture collects and freezes in the small holes of the microphone cap. Throat type microphones should be used for all cold weather operations.

e. Transmitter. - In certain types of transmitters, frequency shifts occur with wide changes in temperature. Consequently the transmitter must be retuned and checked until a relatively stable temperature is reached.

f. Antenna. - Icing is prevalent on all types of antennae. The whip antenna is most satisfactory in this respect and should be used instead of "V" types for radio compasses.

g. Plugs (Jacks). - Cracking occurs on type PL-54. No remedy can be effected.

h. Antenna Shock Mount. - The rubber type shock mounts become very brittle and break in extreme cold weather. A compression type spring can be used for replacement.

11. Carburetor De-icing.

a. Carburetor icing may occur at outside air temperatures of 108^0 C (50^0 F) or less, with humidities greater than 50 percent. Ice formation in the adapter or at the fuel nozzle is indicated by engine roughness and a drop in manifold pressure. Prevention or elimination may be accomplished by raising the carburetor air temperature. A rise in temperature is obtained by closing the intercooler shutters (move control, figure 44-9, to "HOT") or by setting the supercharger controls (figure 40-2) to FULL ON and adjusting power with the throttles. Opening carburetor air filters provides a further increase of carburetor air temperature by drawing heated air from the interior of the wing.

b. Throttle icing at temperature between 0^0 C (32^0 F) with a relative humidity greater than 100 percent is evidenced by a tendency of the throttle to stick. However, this condition is readily alleviated by an increase or decrease of approximately 5.6^0 C (10^0 F) in the carburetor air temperature.

WARNING: Do not exceed the allowable limits for manifold pressure, engine rpm and cylinder head temperatures.

c. Provision is made in some airplanes for installation of alcohol de-icing equipment which injects alcohol into the carburetor air duct between the supercharger and the carburetor inlet. This method of carburetor de-icing should be used:

(1) To start an engine after severe carburetor icing or engine stoppage.

(2) As an aid in determining the cause of power loss or engine roughness. For example, if both de-icing methods (adjusting the engine controls and use of the alcohol system) do not relieve the condition, it can be assumed that carburetor icing is not the cause.

(3) To clear out the engine quickly after the glide at low power through icing conditions.

(4) To obtain full power under icing conditions.

(5) As an alternate method if the use of full turbo or carburetor air filters is prohibited.

RESTRICTED T.O. NO. 01-20EF-1

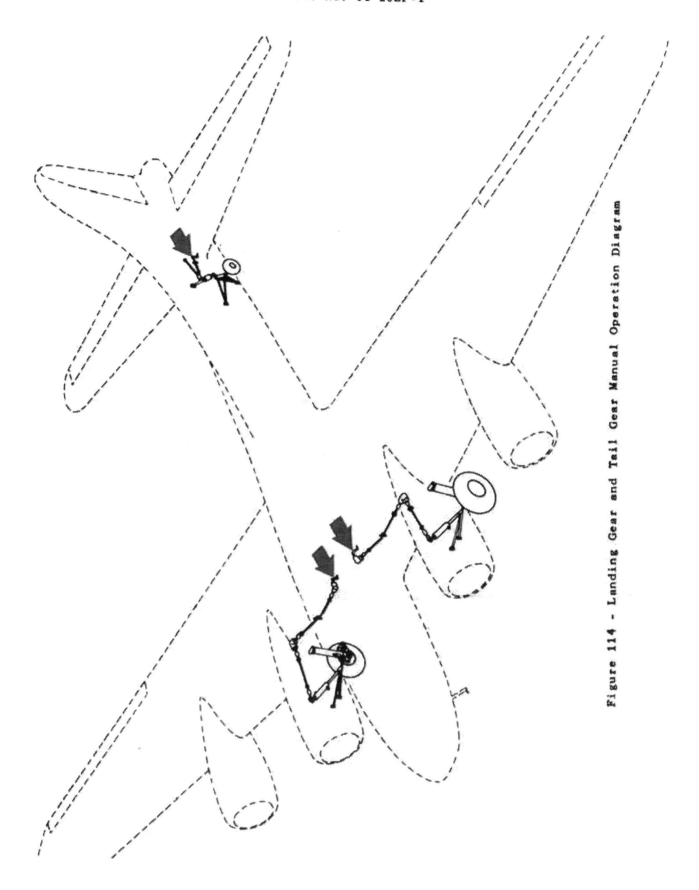

Figure 114 - Landing Gear and Tail Gear Manual Operation Diagram

APPENDIX I

U.S.A. - BRITISH GLOSSARY OF NOMENCLATURE

U.S.A.	BRITISH
Accumulator (hydraulic)	Should not be confused with electrical accumulator or battery
Air field	Aerodrome
Battery (electrical)	Electrical accumulator
Bombardier, bomber	Bomb aimer
Ceiling	Cloud height
Check valve (hydraulic)	Non-return valve
Copilot	Second pilot
Cylinder (hydraulic)	Jack
Dump valve	Jettison valve
Empennage	Tail Unit
Flight indicator	Artificial horizon
Gasoline (gas)	Petrol
Glass, bullet-proof	Armour glass
Gross Weight	All-up weight
Ground (electrical)	Earth
Gyro horizon	Artificial horizon
Gyro pilot	Automatic pilot
(to) Land	(to) Alight
Lean	Weak
Left	Port
(to) Level off	(to) Flatten out
Line, mooring	Mooring guy
Manifold pressure	Boost
Mast, radio	Rod aerial
Overload	Non-standard load
Panel, outboard	Outer plane
Reticle (gun sight)	Graticule
Screen	Filter
Set, command	Pilot controller set
Set, liaison	General purpose set
Ship	Aircraft
Speed, indicated air (IAS)	Air-speed-indicator reading
Stabilizer, horizontal	Tail plane
Stabilizer, vertical	Fin
Stack	Manifold (inlet or exhaust)
Tachometer	Engine speed indicator
Tube (radio)	Valve
Turn indicator	Direction indicator
Valve (fuel or oil)	Cock
Weight empty	Tare
Windshield	Windscreen
Wing	Main plane

APPENDIX II

EMERGENCY OPERATING INSTRUCTIONS

Emergency Operation of Landing Gear.

Manual operation of the main landing wheels is provided for each wheel separately. Hand crank connections are provided in the bomb bay, one to the left of the door in the forward bulkhead, and one to the right. To raise one of the landing wheels, insert the crank (which is clipped to the aft wall of the radio compartment above the transmitter tuning units) into the connection and rotate clockwise. Turn the crank counterclockwise to lower the wheel. (See figure 115.)

Figure 115 - Landing Gear Hand Crank - In Position

DANGER: Be sure the landing gear electric switch is "OFF" before you attempt hand cranking.

Emergency Operation of the Tail Wheel.

The crank that is used for manual operation of the landing wheels is used for manual operation of the tail wheel, bomb bay doors, and wing flaps. Insert the crank into the connection in the tail-wheel compartment (figure 116) and rotate as desired.

Emergency Operation of Bomb Bay Doors.

Insert the hand crank into the torque connection on the step at the forward end of the catwalk in the bomb bay and rotate clockwise to close the doors and counterclockwise to open them. (See figure 117.)

Emergency Operation of Wing Flaps.

Insert the hand crank into the torque connection in the forward end of the camera pit in the radio compartment. Rotate the crank clockwise to lower the flaps and counterclockwise to raise them. (See figure 118.)

5. **Emergency Bomb Release.**

Two emergency release handles are installed, one on the side wall at the pilot's left, and one in the bomb bay at the forward end of the catwalk. The handle in the bomb bay is protected by a guard which may be moved out of position when desired. Pulling of either handle will result in immediate release of bomb door latches, and continued pulling will result in release of all bombs salvo the instant the doors reach the full open position. The bomb bay fuel tanks may also be dropped by an emergency release handle. For instructions on retractions of bomb doors after emergency-release, see figure 119.

6. **Fire Extinguishing Equipment.**

a. General. - Two types of fire extinguishers are used on this airplane, hand fire extinguishers and a CO_2 engine fire extinguisher system.

b. Hand Fire Extinguishers. - Hand fire extinguishers are located as follows: on the side wall of the compartment at the copilot's left, on the forward right side of the aft bulkhead in the upper turret compartment, on the forward bulkhead in the radio compartment, and just aft of the main entrance door.

c. Engine Fire Extinguisher System. - The fire extinguishing system for the four engines is controlled by the copilot by means of a selector valve and two release handles. Operation distributes CO_2 around the engine section through a perforated tube. Each release handle discharges a bottle of CO_2 which is not sufficient to serve more than one engine. Therefore, the selector valve should remain in position for a sufficient length of time after release to fully discharge the bottle. If a fire occurs in another nacelle, reset the selector valve and release the second charge. (See figure 15.) For detailed instructions in case of fire in flight, refer to section II.

7. **Life Raft and First Aid.**

a. Life Raft Controls. - Two automatic ejector life raft installations are provided, one on each side of the enclosure fairing at the rear of the upper turret above the bomb bay. Two control handles are located in the ceiling of the radio compartment. To operate, pull the handles approximately 9 inches.

b. First Aid Kits. - On AAF airplanes serial Nos. 42-5050 and on, first aid kits are located on the bombsight storage box in the navigator's compartment, on the wiring diagram box on the back of the copilot's seat, on the rear bulkhead of the radio compartment, and on the aft side of the bulkhead at the front end of the lower turret compartment.

Figure 116 - Tail Wheel Emergency Handcrank

Figure 117 - Emergency Operation of Bomb Doors

Figure 118 - Emergency Operation of Handcranks

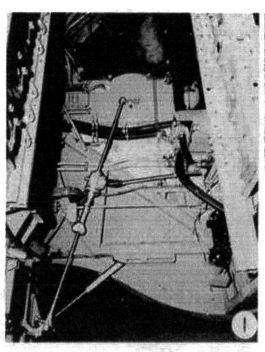
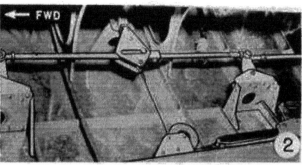

EMERGENCY BOMB RELEASE PROCEDURE

FIRST STEP - Pull either release handle as far as it will go. (See Figure 2) This step will release the doors and allow them to swing open independently of the retracting screws as shown in 1.

SECOND STEP - Pull the release handle again as far as it will go. This step will operate the mechanical salvo and drop all bombs unarmed.

Note: A continuous pull may be maintained on the release handle as the mechanical salvo will occur the instant the doors open far enough to release the interlock on the mechanism. In flight the doors will open so quickly that the entire release operation will appear to be one continuous motion.

DOOR RETRACTION AFTER EMERGENCY RELEASE

FIRST STEP - Observe the emergency release linkage under the hinged door in the floor beneath the pilots' compartment. If the spring has not entirely retrieved the mechanism as shown in 2, complete the resetting of the mechanism by pushing at the hinge of the link as shown in 3.

SECOND STEP - Operate the retracting screws to the fully extended position. This step will engage the latches between the fittings on the screws and the fittings on the door as shown in 4.

THIRD STEP - Retract the doors in the normal manner.

Figure 119 - Emergency Bomb Release Procedure

8. Emergency Exits. (See figure 113.)

a. Navigator's and Bombardier's Compartments. - Through door in aft bulkhead and out the bottom escape hatch (front entrance door, figure 120).

b. Pilot's Compartment.

   (1) Out the bomb bay.

   (2) Out the front entrance door.

c. Upper Turret Compartment. - Out the bomb bay.

d. Radio Compartment.

   (1) Radio Operator. - Out through the bomb bay.

   (2) Auxiliary Crew.

      (a) Out the main entrance door.

      (b) Out the side windows.

e. Ball Turret.

   (1) Out the main entrance door.

   (2) Out either side window.

f. Side Gun Compartment.

   (1) Out the side windows.

   (2) Out the main entrance door.

g. Tail Gunner's Compartment. - Out the side door.

h. Doors and Hatches. - All doors and hatches except the side gunners' windows, which slide forward to open, are quickly releasable. Bomb doors are opened by means of either of the two release handles, one at the left of the pilot and the other at the forward end of the catwalk in the bomb bay.

9. Alarm Bells.

The alarm bell system is used for signals to the crew members in case of an emergency. Three bells are located as follows: one on the bottom side of the navigator's table, one on the wall above the radio operator's table, and one in the tail wheel compartment above the tail wheel boot. (See figure 121.) The bells are controlled by a toggle switch on the pilot's side control panel. Alarm bell signals are usually prearranged between the crew members and will usually differ with each airplane crew.

10. Emergency Operation of Radio Equipment.

a. Operation of Portable Emergency Radio Transmitter (Type SCR-578-A).

   (1) General.

      (a) A complete self-contained portable emergency transmitter is stowed in the aft end of the radio compartment, as illustrated in figure 123, and is provided for operation anywhere away from the airplane. It is primarily designed for use in a small boat or life raft, but it may be placed in operation anywhere a kite can be flown or where water may be found. The unit is usually stowed in the aft end of the radio compartment between the transmitter tuning units and the rear seat. It is equipped with a small parachute to permit dropping from the airplane in event of an emergency.

      (b) When operated, the transmitter emits an MCW signal and is pretuned to the international distress frequency of 500 kc. Automatic transmission of a predetermined signal is provided. Any searching party can "home" on the signal with the aid of a radio compass.

      (c) No receiver is provided.

   (2) Removal from Airplane.

      (a) If the airplane has made an emergency landing on water, the emergency set should be removed at the same time that the life raft is removed. The set is waterproof and will float, and it is not necessary to take any precautions in keeping the equipment out of the water. However, be sure that it does not float out of reach.

Figure 120 - Bottom Escape Hatch - Looking Forward

- 151 -

RESTRICTED

Figure 121

Alarm Bell Tail Wheel Compartment

1. Canopy Emergency Release Handles
2. Left Hand Life Raft Release Handle
3. Right Hand Life Raft Release Handle

Figure 122 - Emergency Controls-Radio Compartment

(b) The emergency set may be dropped from the airplane by use of the parachute attached. The altitude of the airplane when dropping the equipment should be between 300 feet and 500 feet. To drop the equipment, the following steps should be observed:

1. Tie the loose end of the parachute static line to any solid metal structure of the airplane.

CAUTION: Be sure that the static line is in the clear and will not foul.

2. Throw the emergency set out through a convenient opening in the airplane. Parachute will be opened by the static line.

CAUTION: Do not attach static line to any part of one's clothing or body when throwing the equipment through the opening.

(3) Operation. - Complete operating instructions are contained in one of the bags which contain the equipment. Complete instructions for the use of the transmitter are also located on the transmitter itself.

b. Interphone Equipment Failure. - In the event of interphone equipment failure, the audio frequency section of the command transmitter may be substituted for the regular interphone amplifier. To make this connection, the pilot should place his command transmitter control box channel selector switch (figure 4-21) in either channel No. 3 or No. 4 position. Set the interphone jackbox selector switch (figure 42-6) on the "COMMAND" to place the interphone equipment in operation.

NOTE: When the command transmitter control box channel selector switch (figure 4-21) is set in either the No. 3 or No. 4 position for emergency operation of the interphone equipment, it is not possible to establish communication with any station or any other airplane. It is possible at all times to resume normal command set operation by placing the channel selector switch of the command transmitter control box in either the No. 1 or No. 2 position.

c. Substitution of Radio Compass Receiver for Low Frequency Command Set Receiver. - If the low frequency receiver of the command set fails, the radio compass receiver may be substituted, with the pilot having direct control over the compass receiver. To complete this emergency hook-up, the pilot must set his interphone jackbox selector switch (figure 42-6) in the "COMP" position and then place the radio compass selector switch (figure 4-10) in the "ANT" position. The radio compass can then be tuned as desired.

d. Substitution of Liaison Receiver for Low, Medium and/or High Frequency Command Receiver. - In case of the failure of the low, medium and/or high frequency receiver of the command radio equipment, the liaison receiver may be substituted, but the pilot will have only limited control over it. The pilot should first call the radio operator on the interphone system and tell him what frequency he desires to receive, that he is switching the interphone selector switch (figure 42-6) to the "LIAISON" position, and for him (the radio operator) to tune in this frequency and maintain the setting until further advised.

e. Command Set Transmitter Failure. - In case of failure of the command set transmitter, the liaison transmitter may be substituted. The pilot should first call the radio operator on the interphone and have him adjust the liaison transmitter to the frequency he desires to use. He should then set his interphone selector switch (figure 42-6) to the "LIAISON" position and operate his microphone button (figure 44-1) in the same manner that he did when the command set was in operation. When he is through using the liaison transmitter, the pilot should place the interphone selector switch (figure 42-6) in the "INTER" position and tell the radio operator to cut the liaison transmitter off so as to reduce the load on the electrical system.

NOTE: When substituting one receiver for another, such as the compass receiver for the command receiver, the pilot must move his interphone selector switch (figure 42-6) to the "COMMAND" or "LIAISON" position, as the case may be, in order to transmit. At the end of the transmission, he must switch back to the position of the receiver being used. This will have to be done every time that the pilot desires to hold a two-way conversation.

11. Suggested Methods of Abandoning Airplane.

a. General.

(1) Warning Signals. - Three methods enable the pilot to communicate with the crew: the alarm bell system, phone call lamps, and the interphone system. For emergency purposes, the alarm bell should be used in a prescribed manner which is thoroughly understood by all the crew. All signals are given by the pilot. If a commander is present, he will direct the pilot to give the desired signals.

(2) Exits.

(a) Top hatch in radio compartment. Usually used for abandoning the ship after a water landing. Removed by four small release handles on the cover.

(b) Front entrance door. Has an emergency release latch.

(c) Bomb bay. Doors can be opened by either of two emergency releases, one to the left of the pilot and the other in the bomb bay just aft of the door in the forward bulkhead.

(d) Each side gun window.

(e) Main entrance door. Has an emergency release latch.

(f) Tail hatch. Has an emergency release latch.

WARNING: Although rapidity of movement is necessary, it must be remembered that body movement with a parachute is restricted in small quarters. Therefore, care is necessary to avoid bodily injury or damage to the parachute, either of which may prevent bailing out. This applies particularly to head blows.

b. In Flight.

(1) Signal.

(a) Stand by to Abandon: One long ring (approximately 6 seconds).

(b) Abandon: Three short rings (approximately 2 seconds each).

(2) Paths of Egress. (See figure 113.)

(a) Upon the signal to stand by to abandon ship, all members of the crew should fasten on parachutes and prepare to leave the plane through the designated exit.

(b) The bombardier releases and stands by the front entrance door. (See figure 120.)

(c) The navigator stands by the front entrance door. He will first determine position if pilot (or commander) so directs.

(d) The pilot will open the bomb bay doors and release bombs (or tanks) with his release, or instruct either the copilot or another crew member to do so with the other release.

(e) If a gunner is in the upper turret, he will stand by to jump through the bomb bay.

(f) The radio operator will prepare to leave through the bomb bay, first sending any messages designated by pilot (or commander).

(g) The engineer will open and stand by the right side gun window.

(h) The tail gunner will release and stand by the tail hatch door.

(i) Additional crew members will release and stand by main entrance door.

(j) The copilot will prepare to leave through the bomb bay on direction of the pilot (or commander).

(k) The pilot (and commander, if present) will leave through either the bomb bay or the front entrance door.

(3) Switches. - The situation will determine whether fuel and electrical systems should be turned off prior to abandoning ship. Under normal conditions in an area outside of a combat zone, the following switches should be turned off:

(a) Master switch (figure 39-11).

(b) Battery switches (figure 5-16).

(c) Fuel shut-off valves (figure 39-3).

c. Fire in Flight.

(1) In case of engine or wing fires, open the emergency exits. Stand by to abandon: one long ring (approximately 6 seconds).

(2) In case of a cabin fire, the exits should NOT be open. However, it may be desirable to have the crew stand by to abandon without opening the exits. For this the signal is:

Stand by to Abandon - Exits Closed: One short ring (approximately 2 seconds; one long ring (approximately 6 seconds); one short ring (approximately 2 seconds).

d. Crash Landing.

(1) Signal.

(a) Stand by for crash landing: By interphone.

(b) Abandon: Four short rings (approximately 1/2 second each).

(c) Pilot should:

1. Cut engines.
2. Master switch (figure 39-11) - OFF.
3. Battery switches (figure 5-16) - OFF.
4. Fuel shut-off valves (figure 39-3) - OFF.

(2) Egress.

(a) All crew members will take proper stations, remove parachutes, and fasten safety belts upon receiving interphone warning.

(b) At the signal to abandon, all crew members will leave the plane through the most practicable exit. (See figure 113.)

(c) In addition to the seven standard exits, the two side windows in the pilot's compartment are possible exits.

(d) In case some of the exits are blocked by fire, damage or congestion, it may be best to make exit through a rupture in the fuselage if any have occurred. Caution is required in this process to avoid fatal cuts from metal or broken glass.

(e) If there is imminent danger of fire, all personnel should disperse at least 50 feet from the airplane.

e. Landing on Water.

(1) Signal.

(a) Stand by for water landing: By interphone.

(b) Abandon ship: As ordered verbally.

(2) Duties.

(a) At the interphone warning all crew members will fasten on life vests. They will take proper stations and secure safety belts except as noted below.

(b) Pilot (upon direction of commander, if present) will:

1. Release bomb bay tanks if carried and more than half full.

2. Direct bombardier to release all bombs and close bomb bay doors.

3. Direct navigator to determine position.

4. Direct radio operator to send distress signal.

5. Make normal slow landing, flaps down, gear up, contacting water slightly, tail first.

(c) Copilot will:

1. Cut engines, feather propellers, cut all fuel and ignition switches.

NOTE: Engines must be left on until radio and interphone are no longer necessary.

2. Assist pilot as directed.

(d) Bombardier will:

1. Release bombs when directed and close bomb bay doors.

2. Proceed to the radio compartment.

(e) Navigator will:

1. Determine position, notify pilot and radio operator.

2. Proceed to radio compartment and supervise sending of position report.

(f) Radio operator will:

1. Send distress signal as directed by pilot.

2. Send position report as directed by navigator.

(g) Engineer will:

1. See that all crew members have on life vests.

2. Assign a crew member to take along the emergency transmitter when abandoning ship.

3. Open and lock all bulkhead doors.

4. Close all outside doors.

5. See that all other crew members (including tail gunner) are standing by just rear of radio compartment.

6. Stand by radio compartment to open overhead hatch and release life rafts when directed.

Figure 123 - Emergency Radio Transmitter - Stowed

(h) If the airplane is in normal attitude after landing, personnel will abandon through overhead hatch, taking to rafts as follows:

Left Raft	Right Raft
Pilot	Copilot
Bombardier	Navigator
Engineer	Radio Operator
Gunner	Gunner (s)

This assignment may be varied at pilot's direction, under consideration of weight distribution. Water and emergency provisions will be divided equally.

NOTE: Unless plane is sinking rapidly, the rafts will not be cast off until all personnel assigned are aboard.

i. Drill. - Dry run practice on the ground emphasizing the following is of great value:

(1) Signals.
(2) Movement with parachutes on.
(3) Duties.
(4) Taking stations.
(5) Making exit.

12. Emergency Brake Operation.

The emergency system operates the brakes only and pressure is applied through two hand-operated metering valves on the pilot's compartment ceiling. The left-hand lever controls the left wheel and the right-hand lever controls the right wheel. Emergency operation of the brakes is not anticipated unless it is impossible to rebuild the pressure in the service system. If emergency operation becomes necessary, the following procedure is recommended:

a. Manual shut-off valve - CLOSED.

b. Selective check valve - "NORMAL."

c. Determine pressure in emergency accumulator.

CAUTION: Do not attempt to raise the accumulator pressure with the hand pump.

d. Pilot: Operate throttle and rudder.

e. Copilot: Operate emergency brake control.

WARNING: DO NOT "PUMP" EMERGENCY BRAKES. The pressure supply is limited and repeated applications may result in complete loss of emergency braking control.

TAKEOFF- On run-up at full throttle, set and lock turbos at 46 inches. For shortest takeoff use 46 inches, 2500 RPM and one-third flaps and hold three-point position until airplane leaves ground.

CLIMB- Climb at 38 inches, 2300 RPM, auto rich, cowl flaps open, 135 MPH pilot's indicated airspeed. On instrument climbs below 20,000 feet, climb at 160 MPH pilot's indicated airspeed. Use full throttle and set power with turbo regulator. Decrease manifold pressure 1 1/2 inches for each 1000 feet above 30,000 feet.

LEVEL FLIGHT- Use full throttle and set power with turbo regulator at all altitudes. Cowl flaps closed or set to proper cylinder temperature. Mixture auto-rich above 2100 RPM, 30 inches manifold pressure. Auto-lean below 2100 RPM, 30 inches manifold pressure.
NOTE: Do not exceed 30 inches manifold pressure below 2100 RPM.

LONG RANGE CRUISING- For operating conditions refer to long range cruising table below. Set RPM corresponding to gross weight and altitude. Set manifold pressure within ± 1 inch to obtain pilot's indicated airspeed of 155 MPH below 20,000 feet, 150 MPH above 20,000 feet. Use auto-lean mixture. Close cowl flaps or set to obtain proper cylinder temperature. Hold power setting and let airspeed increase as fuel is used. Reset power every three hours to get speed and manifold pressure in the table. For long range climb use above climb instructions.

RULE OF THUMB- Long range cruising 20,000 feet and below. Pilot's indicated airspeed 155 MPH manifold pressure 28 inches ± 1 inch - adjust RPM accordingly.

EMERGENCY OPERATION- High altitude-always use 2500 RPM and manifold pressure as for military power.

LEVEL FLIGHT OPERATION

CONDITION	RPM	MANIFOLD PRESSURE	MIXTURE
TAKEOFF & MILITARY POWER	2500	46 INCHES REDUCE 1 1/2 IN. PER 1000' ABOVE 27,000	AUTO RICH
RATED POWER	2300	38 INCHES REDUCE 1 1/2 IN. PER 1000' ABOVE 30,000	AUTO RICH
MAXIMUM CRUISE	2100	30 INCHES	AUTO LEAN
65 % POWER	2000	29 INCHES	AUTO LEAN
55 % POWER	1900	29 INCHES	AUTO LEAN

NOTE: MANIFOLD PRESSURES ARE APPROXIMATE

LONG RANGE CRUISING

ALTITUDE PILOT'S AIRSPEED	5000 FT 155	10,000 155	15,000 155	20,000 155	25,000 150	30,000 150
GROSS WEIGHT 60,000 LBS	1780 RPM 28 INCHES	1900 28	2030 28	2120 28	2150 31	2220 325
57,000 LBS	1730 RPM 28 INCHES	1850 28	1970 28	2070 28	2120 31	2190 32
55,000 LBS	1680 RPM 28 INCHES	1800 28	1920 28	2020 28	2090 29.5	2150 30.5
52,500 LBS	1630 RPM 28 INCHES	1750 28	1870 28	1970 28	2050 29	2120 30
50,000 LBS	1580 RPM 28 INCHES	1700 28	1810 28	1920 28	2020 28	2080 29.5
47,500 LBS	1530 RPM 28 INCHES	1650 28	1760 28	1860 28	1960 28	2050 29
45,000 LBS	1480 RPM 28 INCHES	1600 28	1710 28	1810 28	1910 28	2000 28
42,500 LBS	1430 RPM 28 INCHES	1550 28	1650 28	1750 28	1850 28	1950 28
40,000 LBS	1400 RPM 28 INCHES	1500 28	1600 28	1700 28	1800 28	1910 28

Figure 124 - Operating Instructions

PILOT'S CONDENSED INSTRUCTIONS—C-1 AUTOPILOT

ENGAGING PROCEDURE

1. **AFTER TAKE-OFF**—Turn on Master Switch.

2. **FIVE MINUTES LATER**—Turn on PDI Switch (and Servo Switch, if separate).

3. **TEN MINUTES AFTER TURNING ON MASTER SWITCH**—*Trim plane mechanically for level flight at cruising speed* ... Bombardier, disengage Autopilot clutch, *center PDI* and lock in place by depressing the Directional Arm Lock. Hold PDI centered until pilot has completed engaging procedure. Then re-engage Autopilot clutch and release Directional Arm Lock.

   ALTERNATE METHOD: Pilot, center PDI by turning airplane in direction of PDI needle; then resume straight-and-level flight.

4. **TO ENGAGE AUTOPILOT**—Put out Aileron Tell-Tale lights with aileron centering knob, then throw on aileron engaging switch. Repeat for rudder, then elevator.

5. **MAKE FINAL AUTOPILOT TRIM CORRECTIONS**—If necessary, use centering knobs to level wings and center PDI.

FOR COLD WEATHER OPERATION, see reverse side.

T. O. 01-20-EF44
AUGUST 20, 1943

OPERATING ADJUSTMENTS

TURN CONTROL
PDI must be centered when Turn Control is used.

AILERON   RUDDER   ELEVATOR

With Autopilot clutch arm at either extreme position, adjust knob for 18° bank.

Adjust knob to center inclinometer ball while in Stabilizer (Bombardier) turns only.

Adjust to obtain constant altitude in either Stabilizer or Turn Control turns.

— Do not use for coordinating Turn Control turns. —

A. R. Screwdriver adjustments below Sensitivity knobs are trimmers for regulating coordination of ailerons and rudders in Turn Control turns only.

CAUTION

Do not adjust mechanical trim tabs while Autopilot is engaged.

TELL-TALE LIGHTS—Must be extinguished in each axis when Autopilot is engaged.

AUTOPILOT TRIM CONTROLS—Before engaging, use to put out Tell-Tale lights. After engaging, use in place of trim tabs for small attitude adjustments. *See note below.*

CONTROLS ALERTNESS OF AUTOPILOT—Adjust clockwise as far as possible until controls chatter; then counterclockwise until chatter stops.

REGULATES AMOUNT OF CONTROL SURFACE ACTION—Adjust clockwise to stop under-control (wallowing). Adjust counterclockwise to stop over-control (hunt).

ADJUSTS COORDINATION OF CONTROL ACTION—Adjust when in maximum Stabilizer (Bombardier) turn, as explained below. Readjustment seldom required.

*NOTE

For large changes in airspeed, CG, or gross weight, disengage Autopilot, retrim with mechanical trim tabs; then re-engage Autopilot.

(Read reverse side before making adjustments)

FLIGHT ADJUSTMENTS AND OPERATION

After the C-1 Autopilot is in operation, the pilot should carefully analyze the action of the airplane to make sure all adjustments have been properly made for smooth, accurate flight control.

When both **TELL-TALE LIGHTS** in any axis are *extinguished*, it is an indication the Autopilot is ready for engaging in that axis.

Each **CENTERING KNOB** is used, before engaging, to adjust the Autopilot control reference point to the straight-and-level-flight position of the corresponding control surface. After engaging, centering knobs are used to make small attitude adjustments.

SENSITIVITY is comparable to a human pilot's reaction time. With sensitivity set *high*, the Autopilot responds quickly to apply a correction for even the slightest deviation. If sensitivity is set *low*, flight deviations must be relatively large before the Autopilot will apply its corrective action.

RATIO is the *amount* of control surface movement applied by the Autopilot in correcting a given deviation. It governs the speed of the airplane's response to corrective Autopilot actions. Proper Ratio adjustment depends on airspeed. If Ratio is too *high*, the Autopilot will *over-control* the airplane and produce a "ship-hunt"; if Ratio is too low, the Autopilot will *under-control*, and flight corrections will be too slow. After Ratio adjustments have been made, centering may require readjustment.

To adjust **TURN COMPENSATION**, have bombardier disengage Autopilot clutch and move engaging knob to extreme right or extreme left. Airplane should bank 18° as indicated by artificial horizon. If not, adjust Aileron Compensation (bank trimmer) to attain 18° bank. Then, if turn is not coordinated, adjust rudder compensation (skid trimmer) to center inclinometer ball. Do not use aileron or rudder compensation knobs to adjust coordination of Turn Control turns.

The **TURN CONTROL** is used by the pilot to turn the airplane while flying under automatic control. To adjust Turn Control, first make sure Turn Compensation adjustments have been properly made, then set Turn Control pointer at beginning of triple-lined area on dial. Airplane should bank 30°, as indicated by artificial horizon. If not, remove cap from aileron trimmer and adjust trimmer until a 30° bank is attained. Then, if turn is not coordinated (inclinometer ball not centered), adjust rudder trimmer to center ball. Make final adjustments with both trimmers and replace caps. Set Turn Control at zero to resume straight-and-level-flight; then recenter.

Never operate Turn Control without first making sure PDI is centered.

The **TURN CONTROL TRANSFER** has no effect unless the installation includes a Remote Turn Control.

The **DASHPOT** on the Stabilizer regulates the amount of "rudder kick" applied by the Autopilot to correct rapid deviations in the turn axis. If a "rudder hunt" develops which cannot be eliminated by adjustment of rudder ratio or sensitivity, the Dashpot may require adjustment. This is accomplished by loosening the lock-nut on the Dashpot, turning the knurled ring up or down until hunting ceases, then tightening the lock nut.

COLD WEATHER OPERATION

When temperatures are between $-12°$ and $0°$ C. ($10°$ and $32°$ F.) Autopilot units must be run for 30 minutes before engaging. If accurate flight control is desired immediately after take-off, this Autopilot "warm-up" should be performed before take-off by turning on the Master Switch during the engine run-up. If warm-up is performed during flight, allow 30 minutes after turning on Master Switch before engaging. When temperatures are below $-12°$C.($10°$F.) units must be preheated for one hour before take-off. Use special heating covers or blankets with heating tubes.

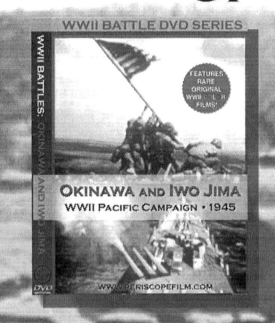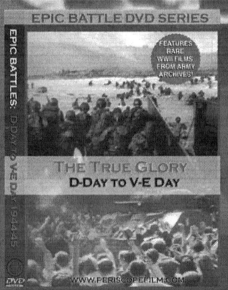

Warships DVD Series

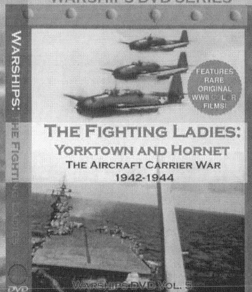

Now Available!

Aircraft At War DVD Series

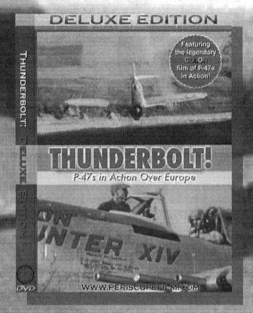
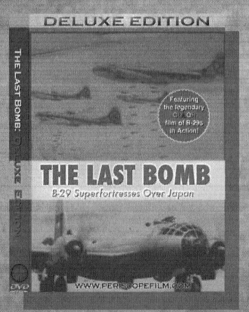
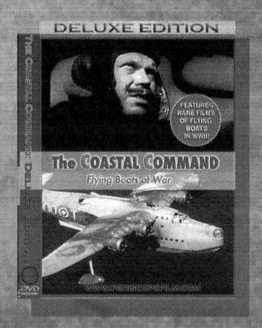

Now Available!

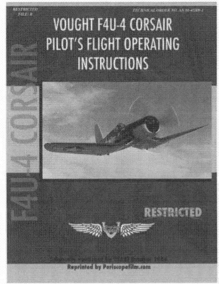
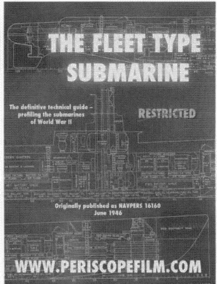
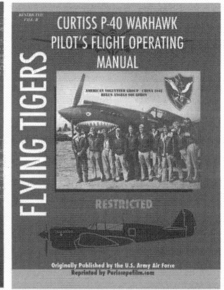
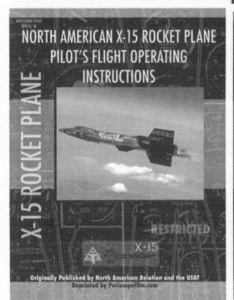
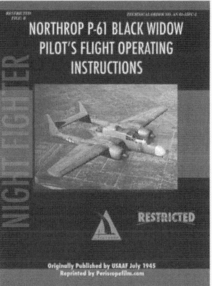
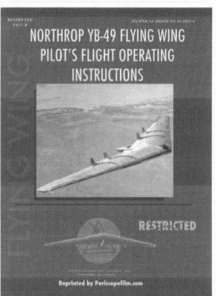
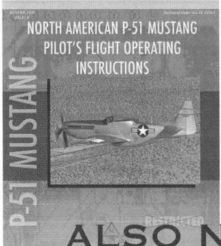

ALSO NOW AVAILABLE FROM PERISCOPEFILM.COM

©2006-2009 Periscope Film LLC
All Rights Reserved
ISBN #978-1-4116-8725-7

Made in the USA
Lexington, KY
21 February 2014